Exoplanets

Planets Beyond The Solar System

Revised, July 2025

Dr Alex Bugeja, PhD

Table of Contents

Introduction

For as long as we have looked to the heavens, we have wondered. Gazing at the countless points of light scattered across the dark expanse, humanity has posed a fundamental question, one that resonates through mythology, philosophy, and science: Are we alone? Is our sun, with its family of planets, a unique occurrence? Or is the cosmos teeming with other worlds, orbiting other stars, perhaps harboring their own forms of life? For millennia, this was a question confined to the realm of speculation. It was the stuff of fiction, a thought experiment for poets and dreamers. Now, within a single generation, that has all changed. We have entered an unprecedented age of discovery, where the query has shifted from "if" to "how many" and "what are they like?"

This book is about that revolution. It is the story of exoplanets—planets beyond our solar system. The term itself is simple, yet it encompasses a menagerie of worlds so vast and varied it challenges our very understanding of what a planet can be. The formal search for these distant worlds was a long and often frustrating endeavor, stretching the limits of our technology and our patience. For decades, astronomers hunted for them, with tantalizing hints but no definitive proof. The breakthrough came not with a whisper, but with a series of discoveries that opened the floodgates.

The first confirmed exoplanets, found in 1992, were not what anyone expected. They were discovered orbiting a pulsar, the incredibly dense, rapidly spinning remnant of a massive star that had gone supernova. While a monumental discovery, it was the confirmation of a planet orbiting a sun-like star in 1995 that truly captured the world's imagination and launched a new field of astronomy. That planet, 51 Pegasi b, was a "hot Jupiter," a gas giant orbiting its star in a blisteringly fast four days. This discovery was a shock, as it defied the then-current theories of planet formation based on our own solar system.

Since that pivotal moment, our catalog of known worlds has exploded. As of mid-2025, astronomers have confirmed the existence of over 5,900 exoplanets in more than 4,400 planetary systems. This number is not static; it grows continuously as new candidates are verified and new missions survey the sky. These are not just blurry points of data; they represent real places. Most of the planets found so far reside in a relatively small patch of our own Milky Way galaxy, yet they hint at a staggering cosmic reality: there are likely more planets than there are stars. The numbers suggest that trillions of planets could populate our galaxy alone.

This book will guide you through this new and thrilling landscape of discovery. We will begin by delving into the history of the search, honoring the pioneers who laid the groundwork for today's planet hunters. We will explore the ingenious methods astronomers have developed to find these worlds, most of which are far too small and dim to be photographed directly. These techniques, such as the transit method, which watches for the faint dimming of a star's light, and the radial velocity method, which detects the gravitational "wobble" of a star, are triumphs of scientific creativity.

Having learned how we find them, we will then journey through the incredible diversity of the planets themselves. Forget the familiar architecture of our own solar system. The galaxy is filled with worlds that were once thought to be impossible. There are gas giants larger than Jupiter orbiting closer to their stars than Mercury does to our sun. There are "Super-Earths," rocky planets significantly more massive than our own, a class of planet that doesn't even exist in our solar system. We have found planets that orbit two stars at once, evoking the iconic double sunset from a galaxy far, far away.

We will investigate planets with surfaces of molten lava and worlds that may be covered entirely by deep, global oceans. Some exoplanets are "rogue planets," untethered to any star, wandering alone through the perpetual darkness of interstellar space. This sheer variety has forced scientists to rethink their models of how planetary systems form and evolve, a process we will examine in detail. Understanding the birth of these worlds helps us understand the story of our own planetary home.

The journey doesn't stop at just finding and categorizing these planets. The true excitement lies in characterizing them—in trying to understand what they are made of and what their conditions are like. We will explore how astronomers are beginning to peer into the atmospheres of these distant worlds. By analyzing the light from a parent star as it passes through a planet's atmosphere, scientists can search for the chemical fingerprints of gases like water vapor, methane, and even carbon dioxide. This is the science of spectroscopy, a tool that allows us to read the "air" of alien worlds from light-years away.

Of course, the ultimate question driving much of this research is the possibility of life beyond Earth. We will dedicate a significant portion of our exploration to the search for habitable worlds. This starts with the concept of the "habitable zone," often called the "Goldilocks zone," the region around a star where conditions might be just right—not too hot, not too cold—for liquid water to exist on a planet's surface. Liquid water, as we know it, is a key ingredient for life.

But the search for habitability is more nuanced than just finding a watery rock. We will consider how a planet's size, atmosphere, and geology all play a role. We'll also look beyond the traditional Goldilocks zone to consider how life might exist in more extreme environments, perhaps on moons orbiting giant planets or on worlds with different chemistries. The search for "biosignatures," the telltale chemical signs of biological processes in a planet's atmosphere, is the next great frontier in this quest.

This revolution in our understanding has been driven by powerful new tools. We will pay special attention to the missions that have made it all possible. NASA's Kepler Space Telescope, launched in 2009, was a game-changer. By staring at a single patch of sky for years, it single-handedly discovered thousands of planets, revealing that small, potentially rocky planets are common in our galaxy. Kepler proved that Earth-sized planets exist around other stars and provided the first real statistics on planet populations.

Now, a new era has dawned with the James Webb Space Telescope (JWST). With its unprecedented sensitivity, JWST is providing breathtaking new insights into the atmospheres of exoplanets, detecting molecules with stunning clarity and giving us our first real glimpse into the chemistry of these worlds. It has successfully made the first thermal emission observations of rocky planets as cool as those in our solar system and has begun to directly image planets, a feat previously reserved for only the most massive, widely separated worlds. This powerful observatory is pushing the boundaries of what we can learn about these distant systems.

Looking ahead, we will preview the next generation of telescopes and missions that promise to continue this torrent of discovery. From ground-based observatories with massive mirrors to future space missions designed specifically to find and characterize Earth-like worlds, the search is only accelerating. We stand on the precipice of answering some of our oldest questions, and perhaps even finding a world that shows the unmistakable signs of life.

This book is a chronicle of that search. It is an exploration of the planets beyond our solar system, from their initial discovery to the cutting-edge science that is revealing their secrets. The story of exoplanets is not just a story about astronomy; it is a story about perspective. For the first time in human history, we can look up at the night sky and know, with scientific certainty, that we are not just looking at stars. We are looking at solar systems. We invite you to join us on this journey to a universe of worlds.

CHAPTER ONE: A Universe of Worlds: An Introduction to Exoplanets

To stand beneath a clear, dark sky is to witness a scene of profound stillness. Countless stars, scattered like diamond dust on black velvet, appear fixed and silent. For most of human history, that stillness was an accepted truth. The stars were distant, unchanging points of light, a beautiful backdrop to our own dynamic world. The only exceptions were the handful of "wandering stars"—the planets of our own solar system—which traced their steady paths against the celestial canvas. The idea that each of those fixed points of light could be a sun in its own right, perhaps with its own family of worlds, was a captivating but unprovable fantasy. Now, we know it is not fantasy, but fact. The quiet sky is, in reality, humming with unseen motion, a cosmic dance of planets orbiting their parent stars.

An exoplanet, or extrasolar planet, is simply a planet located outside of our solar system. Most of the nearly 6,000 confirmed exoplanets orbit other stars, but some are "rogue planets" that wander through the galaxy untethered to any star. This simple definition belies a revolution in thought. It transforms the stars from mere objects of study into places, destinations on a galactic map we are only just beginning to draw. The discovery of these worlds has confirmed that planetary systems are not a special quirk of our own sun, but a common feature of stars throughout the galaxy. We have moved from a universe with eight known planets to one with thousands, and potentially trillions.

The official definition of an exoplanet has evolved as our discoveries have challenged our assumptions. The International Astronomical Union (IAU), the body responsible for cosmic naming conventions, provides a working definition that is more a set of guidelines than a rigid law. According to the IAU, a planet is an object massive enough for its own gravity to have pulled it into a nearly round shape, but not so massive that it ignites nuclear fusion in its core. The upper mass limit for a planet is generally considered to be about 13 times the mass of Jupiter. Beyond that, an object is typically classified as a "brown dwarf," a kind of failed star.

This definition gets a little more complex when considering the relationship between a planet and its star. An additional criterion introduced in 2018 states that for an object to be a planet, it must be significantly less massive than its central star, with a mass ratio of less than about 1-to-25. This helps differentiate a true star-planet system from a binary system where two objects of more comparable mass orbit each other. These technicalities highlight a key point: our understanding of what constitutes a "planet" is being actively shaped and refined by the sheer variety of worlds we are now finding. Nature, it turns out, is far more imaginative than our initial attempts at categorization.

The number of known exoplanets is climbing so rapidly that any precise count is almost immediately out of date. As of the writing of this book, astronomers have confirmed more than 5,900 exoplanets in over 4,400 different star systems. This figure, however, represents only a tiny fraction of the planets believed to exist. These confirmed worlds have been found in a relatively small survey of our own galactic neighborhood. Extrapolating from these findings, scientists estimate that the Milky Way galaxy likely contains a minimum of 100 to 200 billion planets. Some studies suggest the number could be even higher, possibly in the trillions.

The statistical conclusion is staggering: on average, there is at least one planet for every star in our galaxy. Think about that for a moment. Look up at the hazy band of the Milky Way on a dark night. You are not just seeing stars; you are seeing solar systems. The number of planets in our galaxy almost certainly exceeds the number of stars, perhaps by a significant margin. Further estimates suggest that among this cosmic multitude, there could be billions of planets that are roughly the size of Earth. The raw materials for worlds like our own appear to be anything but rare.

These numbers are so vast they can be difficult to grasp. If you were to try to count every one of the estimated 100 billion planets in our galaxy at a rate of one per second, it would take you over 3,000 years. The distances involved are equally mind-bending. The standard unit of measurement for these scales is the light-year, the distance light travels in one year. Traveling at nearly 300,000 kilometers per second, light crosses about 9.5 trillion kilometers in a year. The closest known exoplanet, Proxima Centauri b, is just over four light-years away, a distance of nearly 40 trillion kilometers. Our fastest space probes would take tens of thousands of years to make the journey.

This immense scale is a fundamental challenge in exoplanet science. We cannot, with rare exceptions, "see" these planets directly in the way we can see Mars or Jupiter. They are incredibly small and faint, and the blinding glare of their host stars washes them out completely. Imagine trying to spot a firefly hovering next to a searchlight from hundreds of kilometers away. That is the essence of the problem. As a result, astronomers have had to develop remarkably clever indirect methods to find them, which will be the subject of later chapters.

For centuries, our understanding of planets was based on a single data point: our own solar system. We had small, rocky inner worlds (Mercury, Venus, Earth, Mars) and large, gaseous outer worlds (Jupiter, Saturn, Uranus, Neptune). This neat and orderly arrangement was, for a long time, the only model we had for how a planetary system should look. It was natural to assume that this was the universal template. The discovery of exoplanets has shattered that assumption in the most spectacular way possible.

It turns out our solar system may be something of an oddball. For example, the most common type of planet found so far seems to be a class of world that doesn't even exist in our solar system: the "super-Earth." These are planets with a mass higher than Earth's but substantially below that of our ice giants, Uranus and Neptune. The term refers only to the planet's mass or size, not that it is Earth-like in any other way. These worlds, which can be rocky, ocean-covered, or have dense gas envelopes, appear to be abundant in the galaxy, prompting new questions about why our own solar system lacks one.

Then there are the "hot Jupiters." These are gas giants, similar in size or even larger than our own Jupiter, but they orbit their stars in a breathtakingly close embrace. The first exoplanet found orbiting a sun-like star, 51 Pegasi b, is a classic example. It's a massive planet that completes a full orbit—its "year"—in just four Earth days, orbiting far closer to its star than Mercury does to our sun. The existence of these scorching giants completely upended early theories of planet formation, which held that large planets could only form in the cold outer regions of a star system, like our own Jupiter.

The galactic menagerie doesn't stop there. Astronomers have found "mini-Neptunes," worlds smaller than Neptune but larger than Earth, likely with thick hydrogen and helium atmospheres. There are lava worlds, rocky planets orbiting so close to their star that their surfaces are likely molten oceans of magma. We have found planets that orbit two stars at once, known as circumbinary planets, where one could watch a double sunset reminiscent of a famous science fiction film. The sheer variety is a testament to the diverse outcomes of planetary formation across the cosmos.

There are also planets that belong to no solar system at all. These are the "rogue planets," or free-floating planets, which wander the vast, dark expanse of interstellar space alone. These lonely worlds are not gravitationally bound to any star and may have been ejected from the planetary systems where they formed. While incredibly difficult to detect because they emit no light of their own, scientists believe there could be billions, or even trillions, of these nomadic worlds drifting through the Milky Way, potentially outnumbering the stars themselves.

With thousands of new worlds being logged, a systematic way of naming them became necessary. The convention adopted by the IAU is an extension of the system used for naming multiple-star systems. An exoplanet's official name typically begins with the name of its parent star, which is often a designation from an astronomical catalog (like "HD 189733" or "Kepler-186"). This is followed by a lowercase letter, starting with 'b' for the first planet discovered in that system. Subsequent planets found around the same star are named 'c', 'd', 'e', and so on, in order of their discovery, not their distance from the star. The star itself is considered the implicit 'a' component.

So, when you see a name like "TRAPPIST-1d," it tells you this is the third planet discovered (d) orbiting the star named TRAPPIST-1. While these catalog names can seem dry and technical, they are essential for keeping track of the ever-growing planetary census. In recent years, the IAU has also overseen public campaigns to give official proper names to a selection of exoplanets and their host stars, resulting in more poetic names like "Dimidium" for 51 Pegasi b.

All of this raises a fundamental question: Why do we look? What is the driving force behind this grand endeavor? On one level, the search for exoplanets is about understanding our own origins. By studying how other solar systems form and evolve, we gain a deeper perspective on the history of our own Earth. Seeing the vast range of planetary arrangements helps us understand the specific chain of events that led to our solar system's architecture and, ultimately, to a world capable of supporting life. It helps us place our own existence in a cosmic context.

On another level, it is fueled by one of the most profound questions we can ask: Are we alone? The discovery of worlds orbiting within their star's "habitable zone"—the region where conditions might be right for liquid water to exist on the surface—is a major focus of the search. While finding a planet in this zone is far from a guarantee of life, it is a critical first step. It tells us that the potential abodes for life are not just a theoretical concept but are real places that we can identify and study.

Ultimately, however, the search for exoplanets is also driven by the innate human spirit of exploration. For millennia, we have explored our own planet, pushing into every unknown territory. Now, that impulse is directed outward, toward the cosmos. The pinpricks of light in the night sky are no longer just abstract symbols; they are suns, and we now know they are surrounded by worlds. Each new discovery is a new dot on the map, a new piece of the puzzle. We are living in the first moment in history when we can look up and know, with scientific certainty, that we are part of a vast and crowded universe of worlds.

CHAPTER TWO: The Pioneers: A History of Exoplanet Discovery

The discovery of planets beyond our solar system feels like a modern scientific marvel, a product of the late 20th and early 21st centuries. Yet, the idea itself is ancient, a philosophical seed planted long before the invention of the telescope. The concept that Earth is not unique, that other worlds might orbit other suns, stretches back at least to the ancient Greeks. Thinkers like Epicurus and his fellow atomists argued that an infinite universe must contain an infinite number of worlds. It was a conclusion reached not through observation, but through logic. If the universe was boundless and composed of the same fundamental particles everywhere, it seemed illogical that our world should be a singular creation.

This idea of "cosmic pluralism," however, was largely suppressed for centuries by the prevailing Earth-centered model of the universe established by Aristotle and Ptolemy, which held that Earth was unique. It was not until the Renaissance that the concept was forcefully resurrected. The 16th-century Italian philosopher Giordano Bruno was a fervent supporter of the Copernican model, which placed the Sun at the center of the solar system. But Bruno took it a revolutionary step further. He declared that the universe was infinite, and that the fixed stars were other suns, each with its own retinue of planets, which he called "earths." Bruno suggested these worlds could contain animals and inhabitants, a view that contributed to his condemnation for heresy by the Roman Inquisition and his execution in 1600.

Even after Bruno, the idea of other worlds remained largely in the realm of speculation. Esteemed figures like Isaac Newton considered it a possibility. But wanting to believe in something and proving it scientifically are two entirely different matters. The immense challenge was the sheer scale of the cosmos. Planets are incredibly faint, reflecting only a tiny fraction of their star's light, and their host stars are blindingly bright and unimaginably distant.

For centuries, the technological means to bridge that gap simply did not exist. The search for exoplanets was, for a long time, a search for a ghost in a furnace from a continent away.

The first earnest, though ultimately unsuccessful, attempts to find these worlds began in the 20th century. One of the most famous early efforts centered on Barnard's Star, a dim red dwarf just six light-years from Earth. Starting in the 1960s, Dutch-American astronomer Peter van de Kamp, working at Swarthmore College's Sproul Observatory, began announcing the discovery of planets orbiting this nearby star. He used a technique called astrometry, which involves meticulously measuring the precise position of a star over many years, looking for a tell-tale "wobble" caused by the gravitational tug of an orbiting companion.

Van de Kamp's observations, taken over decades, suggested that Barnard's Star was indeed wobbling. He initially claimed to have found a gas giant planet, and later revised his claim to a two-planet system. For about a decade, many in the astronomical community accepted his findings, and the idea of planets around Barnard's Star captured the public imagination. However, other astronomers could not replicate his results. Eventually, it was shown that the "wobbles" van de Kamp had detected were not from planets, but were systematic errors likely caused by periodic adjustments and upgrades to the observatory's telescope lens. Van de Kamp, however, went to his grave in 1995 still believing he had discovered the first exoplanets.

The story of Barnard's Star served as a cautionary tale for the emerging field of exoplanet hunting. It highlighted the immense difficulty of the measurements required and the ease with which systematic errors could masquerade as discoveries. This history of false starts made the community deeply skeptical of new claims, creating a high bar for any future announcement. The hunt was not for the faint of heart, and the road to the first confirmed discovery was littered with tantalizing but unverified signals.

One such signal came in 1988, from a Canadian team led by astronomers Bruce Campbell, Gordon Walker, and Stephenson

Yang. They were monitoring the star Gamma Cephei and detected a periodic wobble consistent with a planet. However, the data was not quite clean enough to make a definitive claim, and they retracted it in 1992, concerned that the star's own activity might be mimicking a planet's signal. In a twist of history, their initial suspicion was correct. Years later, in 2003, with better data and techniques, the planet Gamma Cephei Ab was officially confirmed, making it technically the first exoplanet to be discovered, lost, and then found again.

A similar near-miss occurred around the same time. In 1989, a team led by David Latham at the Harvard-Smithsonian Center for Astrophysics reported an unseen companion around the star HD 114762. They were cautious in their paper, noting that the object could be a planet but was more likely a brown dwarf, an object more massive than a planet but not massive enough to be a star. Their hesitancy, born from the field's history of premature claims, meant the discovery didn't initially register as the first planet. Later analysis confirmed the object's mass places it in the brown dwarf category, not the planetary one. These "almost-discoveries" show that astronomers were hovering on the edge of a breakthrough, their instruments and methods just reaching the necessary precision.

The first undisputed, confirmed discovery of planets outside our solar system finally came in 1992, and it arrived from a completely unexpected direction. The discovery was made not around a stable, sun-like star, but around a pulsar. A pulsar is the tiny, incredibly dense, rapidly spinning corpse of a massive star that has exploded in a supernova. They are cosmic lighthouses, sweeping a beam of radio waves across space with a regularity so precise they can rival atomic clocks.

Polish astronomer Aleksander Wolszczan had detected a new millisecond pulsar, designated PSR B1257+12, in 1990 using the Arecibo radio telescope in Puerto Rico. As he and his colleague Dale Frail began timing the radio pulses from this object, they noticed something odd. The pulses weren't perfectly regular; they arrived slightly early or slightly late in a complex but repeating

16

pattern. This indicated that the pulsar itself was moving, being tugged back and forth by the gravity of unseen companions.

After painstakingly ruling out all other possibilities, Wolszczan and Frail published their stunning conclusion in January 1992: the pulsar was orbited by two planets. These were not the gas giants astronomers had expected to find, but small, rocky worlds with masses a few times that of Earth. A third, even smaller planet, with a mass comparable to our Moon, was confirmed a couple of years later. The discovery was monumental, providing the first rock-solid proof that planets existed beyond our solar system.

Yet, the discovery also posed a profound riddle. A supernova is an event of unimaginable violence, an explosion that should have obliterated any pre-existing planets. Finding worlds orbiting a pulsar was like discovering children playing in the pristine ruins of a demolished building. It suggested that these planets were not survivors, but rather "second-generation" worlds, formed out of the debris disk left behind after the star's explosive death. While a landmark achievement, the pulsar planets were so strange, their environment so alien and inhospitable, that the quest to find a planet around a star like our own Sun continued with renewed vigor.

That ultimate prize was finally claimed just three years later. In 1995, Swiss astronomers Michel Mayor and his PhD student, Didier Queloz, from the University of Geneva were conducting a survey of 142 sun-like stars at the Haute-Provence Observatory in southern France. They were using a new, highly sensitive spectrograph named ELODIE, designed to detect the subtle wobble in a star's motion caused by an orbiting planet. One of their targets was a non-descript star, 51 Pegasi, located about 50 light-years away.

As Queloz analyzed the data from 51 Pegasi, he saw something that at first seemed unbelievable. The star was wobbling, and it was wobbling dramatically. The signal was strong and clear, but the period of the wobble was shocking. It indicated a planet at least half the mass of Jupiter was whipping around its star every

4.2 days. This was utterly bizarre. According to the planetary formation theories of the time, a gas giant like Jupiter could only form in the cold, distant regions of a solar system, where it would take a decade or more to complete an orbit, not just a few days.

The result was so strange that Mayor and Queloz were initially skeptical, worrying it was an instrumental error or some kind of stellar pulsation they didn't understand. But after re-observing the star and checking their data, the signal remained undeniable. They had found a planet, but it was a type of planet no one had predicted: a "hot Jupiter." Realizing they had a major discovery on their hands and that American teams were hot on their heels, they rushed to publish their findings.

On October 6, 1995, Mayor and Queloz announced their discovery at an astronomy conference in Florence, Italy. The news was met with a mixture of excitement and disbelief. The discovery of the first planet orbiting a sun-like star was a watershed moment in the history of astronomy, a discovery for which they would share the 2019 Nobel Prize in Physics. The announcement sent a ripple through observatories around the world. Within days, American astronomers Geoffrey Marcy and Paul Butler, who had been conducting their own search, were able to point their telescope at 51 Pegasi and confirm the planet's existence. The race was over, and a new era of exploration had begun.

The discovery of 51 Pegasi b did more than just prove that planets around normal stars existed; it shattered the prevailing models of how planetary systems were supposed to behave. It demonstrated that our own solar system, with its neat division of small rocky inner planets and large gassy outer planets, was not the only blueprint. Nature, it turned out, was far more creative and chaotic than we had imagined. The discovery opened the floodgates, and in the years that followed, the trickle of planetary candidates turned into a torrent, confirming that we live not in a lonely solar system, but in a galaxy full of worlds.

CHAPTER THREE: Finding Shadows: The Transit Method of Detection

Of all the ingenious techniques astronomers have devised to hunt for worlds orbiting distant stars, none has been more prolific than the transit method. In principle, it is beautifully simple. Imagine sitting in a park at night, watching a bright, distant streetlamp. If a tiny moth were to fly directly between you and the lamp, it would block a minuscule fraction of the light, causing a slight, momentary dip in the lamp's brightness. This is the essence of the transit method. When a planet's orbit carries it directly in front of its star from our point of view, it casts a shadow toward us, causing a brief and periodic dip in the star's observed brightness.

This simple concept, however, masks a monumental challenge. The effect is incredibly subtle. A gas giant the size of Jupiter crossing a Sun-like star blocks only about one percent of the star's light. For an Earth-sized planet, the dip is a hundred times smaller, a mere 0.01 percent dimming. Detecting such a tiny flicker from a star many light-years away requires extraordinarily precise measurement of starlight, a practice known as photometry. Astronomers must be able to measure the brightness of a star with a precision of a few parts per million, a feat made possible by modern sensitive electronic detectors called CCDs, the same technology found in digital cameras.

The data gathered from these observations is plotted on a graph called a light curve, which shows the star's brightness over time. A planet transit appears as a characteristic dip in this curve. The curve is normally a flat line, indicating the star's steady brightness. As the planet begins to cross the edge of the star, an event called ingress, the light curve starts to dip downwards. While the planet is fully in front of the star, the curve flattens out at the bottom. As the planet begins to leave the other side of the star, an event called egress, the light curve slopes back up to its normal level.

This simple shape holds a treasure trove of information. The most fundamental discovery, of course, is the existence of the planet itself. But the light curve tells us much more. The depth of the dip—how much the starlight dims—is directly related to the ratio of the planet's size to the star's size. Specifically, the percentage of light blocked is proportional to the area of the planet's disk divided by the area of the star's disk. Since we can usually estimate the size of the star based on its color and brightness, measuring the transit depth allows astronomers to calculate the physical radius of the planet with remarkable accuracy. This is the only method that directly measures a planet's size.

The timing of the transits is equally revealing. By measuring the time between consecutive dips, astronomers can determine the planet's orbital period—the length of its year. Once the period is known, a simple but powerful law of physics, Kepler's Third Law of Planetary Motion, can be applied. This law relates the orbital period of a planet to its average distance from its star. So, by timing the transits, we can calculate how far the planet orbits from its parent star. Knowing this distance, combined with knowledge of the star's temperature, allows scientists to make a first-order estimate of the planet's surface temperature, a key factor when considering its potential for habitability.

However, the transit method has a major limitation: geometry. For a transit to be observable, the planet's orbital plane must be almost perfectly aligned with our line of sight. If the orbit is tilted even slightly, the planet will pass above or below its star from our perspective, and no transit will occur. The odds of this alignment depend on the size of the orbit; the closer a planet is to its star, the higher the probability of a transit. For a hot Jupiter in a tight orbit, the chances might be around 10%. But for a planet like Earth, orbiting a Sun-like star at a similar distance, the probability of a chance alignment is less than half a percent.

This means that for every transiting planet we find, there are hundreds more orbiting at angles that make them invisible to this method. The transit technique can't tell us if a particular star has planets, but by observing a huge number of stars, it can give us a powerful statistical understanding of how common planets are. It is a numbers game, and to win it, you need to stare at a lot of stars for a long time. This is precisely why the most successful transit surveys have been conducted by dedicated space telescopes.

The first of these was the French-led CoRoT mission, but the true game-changer was NASA's Kepler Space Telescope, launched in 2009. Kepler was essentially a giant digital camera designed for one purpose: to hunt for transiting planets. It stared relentlessly at a single, crowded patch of sky in the constellation Cygnus, continuously monitoring the brightness of about 150,000 stars for four years. Its location in space, free from the blurring and day-night interruptions of Earth's atmosphere, allowed it to achieve unprecedented photometric precision. Kepler's patient vigil revealed thousands of planetary candidates, proving that planets, and especially small, rocky planets, are incredibly common throughout the galaxy.

Kepler's successor, the Transiting Exoplanet Survey Satellite (TESS), launched in 2018, employs a different strategy. Instead of staring deeply at one patch of sky, TESS is scanning almost the entire sky, searching for transiting planets around the nearest and brightest stars. The planets TESS finds are ideal candidates for follow-up studies. Because their host stars are bright, it is easier for other telescopes, like the James Webb Space Telescope, to perform the detailed observations needed to characterize their atmospheres. Alongside these space missions, numerous ground-based surveys like HATNet and SuperWASP have also successfully found hundreds of planets, typically focusing on finding larger "hot Jupiters" whose deeper transits are more easily detectable through the Earth's atmosphere.

Of course, not every dip in a star's light is a planet. The transit method is plagued by the problem of "false positives," astrophysical phenomena that can mimic a planetary transit. The most common culprit is an eclipsing binary star system. Sometimes, a faint pair of stars orbiting each other can be located in the background, directly behind the brighter target star. When one of these background stars eclipses the other, their combined light, blended with the light of the foreground star, creates a small dip that can look just like a planet transit. Another common false positive is a "grazing" eclipsing binary, where a stellar companion to the target star only partially blocks its light, producing a shallow, V-shaped dip instead of the flat-bottomed, U-shape typical of a full planetary transit.

Astronomers have developed several techniques to weed out these impostors. The shape of the light curve itself is a primary clue. A transit by a planet, which is much smaller than its star, tends to have steep sides and a flat bottom, often described as "U-shaped" or "bucket-shaped". An eclipse by another star, a much larger object, often produces a more "V-shaped" dip because the ingress and egress take much longer relative to the total duration. However, this is not a foolproof indicator.

To confirm a planetary candidate, astronomers require more evidence. First, the dips must be periodic. Seeing a single dip is not enough; at least three transits are needed to confidently establish a repeating pattern and rule out random stellar fluctuations. Second, follow-up observations are crucial. Often, this involves using the radial velocity method, which will be discussed in the next chapter. If a transit signal is accompanied by a corresponding gravitational wobble from the star, the case for a planet becomes much stronger.

This combination of the transit and radial velocity methods is particularly powerful. As we've seen, the transit method gives us the planet's radius. The radial velocity method, on the other hand, gives us its mass. With both the radius (from the transit) and the mass (from the wobble), astronomers can calculate one of the most fundamental properties of a planet: its density. Density is our first and most important clue to a planet's bulk composition. A high density suggests a rocky world, like Earth or Mars. A low density implies a gas giant, rich in hydrogen and helium, like Jupiter or Saturn. This synergy between the two methods turns a simple shadow into a world with tangible physical properties.

The transit method can reveal even more subtle details. The star's light doesn't just get blocked by the solid body of the planet; a tiny fraction of it filters through the planet's atmosphere on its way to our telescopes. As this light passes through the atmosphere, different molecules absorb light at specific wavelengths, or colors. By observing the transit in many different colors of light, a technique known as transmission spectroscopy, astronomers can look for the tell-tale absorption signatures of various gases. This allows us to begin to analyze the chemical composition of an alien world's air, searching for molecules like water, methane, and carbon dioxide, a topic we will explore in much greater detail later.

In systems with multiple planets, the transit method can unlock another layer of discovery through a technique called Transit Timing Variations, or TTV. If a star hosts only one transiting planet, its transits should occur with clockwork regularity. However, if other planets are present in the system—even ones that do not transit—their gravitational pull will tug on the transiting planet, causing it to arrive slightly early or late for its transit appointments. These tiny variations from a perfectly periodic schedule, sometimes just a few minutes, are a direct measurement of the gravitational interactions between the planets.

By carefully modeling these timing variations, astronomers can not only deduce the presence of other, non-transiting planets in the system but can also determine the masses of all the interacting planets. This is a remarkable feat; it allows us to weigh planets using only the light from their star, without needing the separate, time-consuming observations required for the radial velocity method. The first confirmed detection of a planet via TTV was Kepler-19c, a non-transiting world whose presence was inferred solely from its gravitational influence on the transiting planet Kepler-19b. The TTV method is especially sensitive in compact, resonant systems, where the planets' orbital periods are in simple mathematical ratios, allowing their gravitational nudges to build up over time.

From a simple, repeating shadow, astronomers can thus deduce a planet's existence, its size, the length of its year, its distance from its star, its temperature, its density, the composition of its atmosphere, and even the presence and masses of its hidden siblings. It is a testament to scientific ingenuity that so much can be learned from such a tiny flicker of light. The transit method has been responsible for the vast majority of exoplanet discoveries to date, transforming our view of the cosmos from a lonely solar system into a galaxy teeming with worlds. It has turned stars into solar systems and points of light into places.

CHAPTER FOUR: The Wobbling Star: The Radial Velocity Method

While the transit method hunts for the shadow of a planet, a sister technique, known as the radial velocity method, searches for its substance. It doesn't look for the planet itself, but for the gravitational ghost it impresses upon its parent star. Every planet, no matter how small, exerts a gravitational pull on the star it orbits. This pull is, of course, minuscule compared to the star's own immense gravity, but it is not zero. The common picture of a planet serenely circling a stationary star is a convenient simplification. In reality, a planet and its star are dance partners, and they orbit a mutual point of gravity between them called the barycenter, or center of mass.

Imagine a hammer thrower whirling a heavy weight around them before a throw. The athlete doesn't stay perfectly still; they lean back and shuffle their feet in a small circle to counterbalance the force of the spinning weight. The star and planet do the same thing. Because the star is vastly more massive, the center of mass for the system is usually located deep inside the star itself, but it is not at its exact center. As the planet traces its wide orbit, the star is forced into its own much smaller, corresponding orbit around this barycenter. From light-years away, this motion appears to us as a slight "wobble." For decades before the first confirmed discovery, astronomers knew that if they could just measure this stellar wobble, they could infer the presence of an unseen planet.

The challenge is that this wobble is impossibly small to see directly. Trying to visually track such a tiny change in a star's position is a technique called astrometry, and it has only recently begun to bear fruit after decades of effort. The radial velocity method, also known as Doppler spectroscopy, sidesteps this by not trying to see the star move side-to-side, but by measuring its velocity as it moves towards and away from us. This is where one of the most familiar concepts in physics comes into play: the Doppler effect. It's the same phenomenon that causes the pitch of an ambulance siren to rise as it races towards you and drop as it speeds away. The sound waves are compressed as the source approaches, leading to a higher frequency (higher pitch), and stretched as it recedes, leading to a lower frequency (lower pitch).

Light behaves in the same way. If a light-emitting object like a star is moving towards an observer, the light waves it emits are compressed, shifting their color towards the blue end of the spectrum. This is called a "blueshift." If the star is moving away, its light waves are stretched, shifting them towards the red end of the spectrum—a "redshift." As a planet's gravity pulls its star through a tiny orbit, the star will, from our perspective, be moving towards us for half of its "wobble" and away from us for the other half. By measuring these periodic shifts in the starlight's color, astronomers can detect the planet's gravitational influence.

To measure such subtle shifts requires an instrument of incredible precision: a high-resolution spectrograph. A spectrograph acts like a very sophisticated prism, taking the combined light from a star and splitting it into its constituent colors, producing a spectrum. This stellar spectrum is not a perfect, unbroken rainbow. It is crossed by thousands of dark, narrow vertical lines known as absorption lines or Fraunhofer lines. Each line is a chemical fingerprint, a specific wavelength of light that has been absorbed by elements like hydrogen, sodium, iron, and calcium in the star's cooler outer atmosphere. These lines provide a precise "barcode" for the star's light.

When the star wobbles towards us, this entire barcode—every single spectral line—shifts slightly towards the blue. When it wobbles away, the whole pattern shifts to the red. The spectrograph's job is to measure these shifts with phenomenal accuracy. The planet hunters who pioneered this method were not looking for a dramatic change in color, but a minuscule back-and-forth slide of this spectral barcode. Modern instruments can detect velocity changes of less than one meter per second—the speed of a slow walk—in a star hundreds of light-years away. It's a measurement feat equivalent to noticing a single grain of sand being added to a large truck.

When astronomers make these measurements over time, they can plot them on a graph called a radial velocity curve. This graph shows the star's line-of-sight velocity on the y-axis and time on the x-axis. If there is no massive companion, the data points will scatter randomly around zero. But if a planet is present, a distinct, repeating pattern emerges. The star's velocity will cyclically increase (moving away, redshifted) and decrease (moving toward, blueshifted). The result is a wave-like curve that holds the secrets of the unseen planet's orbit.

The time it takes for the wave to repeat, from one peak to the next, reveals the orbital period of the planet. If the curve repeats every four days, like that of 51 Pegasi b, then the planet's "year" is four days long. The amplitude of the curve—the height of its peaks and the depth of its troughs—tells astronomers the maximum speed of the star's wobble. This speed is directly related to the planet's mass and its proximity to the star. A more massive planet, or a planet orbiting very close to its star, will exert a stronger gravitational tug, creating a larger wobble and a higher-amplitude velocity curve. Knowing the star's mass (which can be estimated from its spectral type and brightness), astronomers can use the amplitude and period to calculate the planet's mass.

The radial velocity curve can also reveal the shape of a planet's orbit. If a planet has a perfectly circular orbit, the star's wobble will be smooth and uniform, producing a perfectly symmetrical wave known as a sine curve. However, many planets have elliptical, or eccentric, orbits. According to Kepler's Second Law of planetary motion, a planet moves fastest when it is closest to its star (at a point called periastron) and slowest when it is farthest away (apoastron). This changing speed gives the star a correspondingly non-uniform wobble. Instead of a smooth, symmetrical push and pull, the star gets a short, sharp tug as the planet whips by at its closest approach, followed by a long, gentler pull as the planet moves through the far side of its orbit. This results in an asymmetric, skewed radial velocity curve. By analyzing the precise shape of this skewed wave, astronomers can calculate the eccentricity of the planet's orbit.

The discovery that many exoplanets, particularly the early hot Jupiters, had highly eccentric orbits was another major shock to planetary science. In our solar system, the large planets have nearly circular orbits. The prevalence of eccentric orbits elsewhere suggested that planetary systems could be much more dynamically chaotic places than our own, perhaps shaped by violent planet-planet scattering events in their past.

Despite its power, the radial velocity method has a fundamental limitation, often called the "sin i problem." The technique only measures the star's motion along our line of sight—its *radial* velocity. It tells us nothing about the star's motion perpendicular to our view. The magnitude of the wobble we measure depends entirely on the orientation of the planetary system in space. Imagine a merry-go-round. If you view it from the side (an "edge-on" view, with an orbital inclination i of 90 degrees), you see its full back-and-forth motion. But if you view it from directly above (a "face-on" view, with an inclination of 0 degrees), you see it spinning, but you detect no motion towards or away from you.

The same is true for a star-planet system. If the orbit happens to be aligned perfectly edge-on to our line of sight, we measure the star's true velocity and can calculate the planet's true mass. If the system is face-on, the star wobbles in a plane perpendicular to our view, producing no Doppler shift at all, making the planet completely invisible to this method. Since in most cases we don't know the orbital inclination, we cannot determine the planet's true mass. What we can calculate is a lower limit, or *minimum mass*. This value is reported as "M sin i," where 'M' is the true mass and 'i' is the unknown inclination angle. Since sin(i) is always between 0 and 1, the minimum mass is always less than or equal to the true mass. For most systems, statistical arguments suggest the true mass is likely to be reasonably close to the minimum mass, but there will always be a small chance that we are viewing a system at a low inclination and are detecting a very massive planet or even a brown dwarf masquerading as a smaller planet.

This limitation also creates a natural bias in the method. The planets that are easiest to detect are those that produce the largest radial velocity signals. These are the most massive planets orbiting closest to their stars. A Jupiter-mass planet in a tight, days-long orbit can induce a stellar wobble of hundreds of meters per second. By contrast, an Earth-like planet in a one-year orbit around a Sun-like star produces a wobble of only about 9 centimeters per second. Finding such a tiny signal buried in the data is an immense challenge. This observational bias is the main reason why the first planets discovered were "hot Jupiters"—not necessarily because they are the most common type of planet in the galaxy, but because they were the easiest to find with the technology of the time.

Another major difficulty is the star itself. Stars are not the perfect, stable spheres of light we might imagine. They are roiling, active balls of plasma. Dark, cool starspots and bright, hot patches called plages rotate across the stellar surface, creating distortions in the spectral lines that can mimic the Doppler shift of a planet. The surface of the star can also bubble and churn with convective motions, and the entire star can pulsate, all of which introduce "jitter" or astrophysical noise into the radial velocity data. For very active, young stars, this stellar noise can create signals equivalent to tens of meters per second, completely washing out the signal of all but the most massive planets. Teasing a faint planetary signal out from the star's own intrinsic variability is one of the most significant challenges in the field, requiring complex modeling and simultaneous monitoring of stellar activity indicators.

The instruments that perform this delicate work are masterpieces of engineering. The discovery of 51 Pegasi b was made with the ELODIE spectrograph in France. Its successors have pushed the boundaries of precision ever further. Key instruments include the High Resolution Echelle Spectrometer (HIRES) at the Keck Observatory in Hawaii and the High Accuracy Radial velocity Planet Searcher (HARPS) at the La Silla Observatory in Chile. To achieve their incredible stability, these spectrographs are often sealed in vacuum chambers and temperature-controlled rooms to isolate them from any environmental changes that could alter their measurements. The development of these ultra-stable spectrographs has been the driving force behind the success of the radial velocity method.

For almost fifteen years, the wobble method reigned as the king of planet detection. While the transit method has since overtaken it in terms of the sheer number of planets found, the radial velocity technique remains indispensable. Its greatest strength today lies in its synergy with the transit method. When a planet is observed to transit, we know its orbit must be very close to edge-on, meaning the inclination i is near 90 degrees and sin(i) is almost exactly 1. In these cases, the $m \sin(i)$ problem vanishes. The radial velocity measurement gives us the planet's true mass, while the transit measurement gives us its physical radius. With both mass and radius in hand, astronomers can calculate the planet's bulk density—the single most important clue to its composition, telling us whether we have found a rocky super-Earth or a puffy gas giant. This powerful combination turns a stellar wobble and a passing shadow into a tangible world.

────────────────────────────

CHAPTER FIVE: Bending Light: Gravitational Microlensing

In the toolkit of the planet hunter, the transit and radial velocity methods are the reliable workhorses, patiently watching for recurring signals. Gravitational microlensing, by contrast, is the wild card. It is a technique rooted in one of the most profound and mind-bending concepts in all of physics, Albert Einstein's general theory of relativity, and it hunts for planets through a fleeting and magnificent cosmic coincidence. It does not rely on light from the planet or even its host star. Instead, it uses the light from a much more distant, unrelated star as a beacon, watching for the moment when a planetary system drifts in front of it, its gravity acting as a lens to magnify the background light.

Einstein's theory famously describes gravity not as a force, but as a curvature of spacetime. Any object with mass creates a dent in the fabric of spacetime, much like a bowling ball placed on a trampoline. Light, as it travels through the cosmos, follows these curves. When light from a very distant object passes by a massive object in the foreground, its path is bent. This phenomenon, known as gravitational lensing, was first confirmed by Sir Arthur Eddington in 1919, who observed that starlight passing near the Sun during a solar eclipse was indeed deflected, providing the first empirical proof of general relativity.

On a grand scale, massive galaxy clusters can act as "strong lenses," bending light from background galaxies so dramatically that they appear as distorted arcs, streaks, or even multiple distinct images. Gravitational microlensing is the same physical effect but on a much smaller scale, where the lensing object is a single star or planet. In this case, the separation between the lensed images is minuscule—typically on the order of a milliarcsecond, an angle too small to be resolved by any current telescope. Instead of seeing multiple images, we observe a different effect: the foreground star's gravity focuses the light from the background star, causing a temporary and significant increase in its apparent brightness. The foreground star becomes a cosmic magnifying glass.

For this to happen, the alignment between the observer (on Earth), the foreground "lens" star, and the background "source" star must be nearly perfect. As the lens star drifts across our line of sight to the source star, the source star will appear to brighten smoothly and then fade away symmetrically as the alignment ends. This characteristic brightening, plotted on a light curve, is a microlensing event. The entire phenomenon is transient, typically lasting for a few weeks or months, and because it relies on the chance alignment of moving stars, it is a one-time-only event. Once it's over, it will not repeat.

This presents both a challenge and an opportunity. The rarity of these alignments means astronomers must monitor an enormous number of stars simultaneously to catch an event in progress. The best place to look is toward the most crowded parts of the sky, such as the galactic bulge, the dense central region of the Milky Way. Surveys like the Optical Gravitational Lensing Experiment (OGLE) and Microlensing Observations in Astrophysics (MOA) do just this, pointing their telescopes at millions of stars every night, searching for these tell-tale brightenings.

The shape and duration of the brightening curve tell astronomers about the lensing object. A more massive lens star creates a stronger gravitational field and a longer-lasting event. The peak magnification, which can make the source star appear hundreds of times brighter, depends on how perfectly the objects align. But the true magic for planet hunters happens when the lens star is not alone. If the star hosts a planet, that planet has its own, much smaller gravitational field. This adds a secondary, tiny "dent" to the main spacetime curvature created by the star.

If the background source star happens to pass behind this planetary perturbation, the planet itself acts as a small, secondary lens. This creates a brief, sharp anomaly on the otherwise smooth, bell-shaped light curve of the main stellar lensing event. This planetary signal typically appears as a short-lived spike or sometimes a dip, lasting anywhere from a few hours to a couple of days. It's like finding a small chip in the cosmic magnifying glass, one that reveals the presence of an unseen world. Detecting this sharp little flicker atop the broader wave of the stellar event is the signature of a microlensed planet.

This technique is powerful because the planetary signal provides a direct measurement of the ratio between the planet's mass and its host star's mass. The duration and shape of the small spike tell astronomers about the planet's mass relative to its star, while its position on the main light curve reveals the planet's projected separation from the star at that specific moment in time. It provides a "snapshot" of the system. While this doesn't yield the planet's full orbit like the radial velocity method, it gives a direct handle on the mass ratio, a key physical parameter.

Gravitational microlensing has a unique set of strengths that make it a crucial and complementary tool to other detection methods. Its most significant advantage is its remarkable sensitivity to planets of very low mass. While the transit and radial velocity methods struggle to detect the tiny signals from Earth-mass planets, the gravitational disturbance from a small world can still produce a sharp, clear spike in a microlensing light curve. In January 2006, this method was used to find OGLE-2005-BLG-390Lb, a planet with a mass only about five times that of Earth, which at the time was the lowest-mass planet ever detected.

Furthermore, microlensing is most sensitive to planets orbiting at greater distances from their stars, typically in orbits of a few Astronomical Units (AU), the region where planets like Mars, Jupiter, and Saturn reside in our own solar system. This is because the planet is most likely to cause a detectable anomaly if it is located near the "Einstein radius" of the lens star—the radius of the ring of light that would be seen if the alignment were perfect. For a typical star in our galaxy, this radius corresponds to a physical distance of a few AU. This makes microlensing perfectly suited to probe the cold, outer regions of planetary systems, a crucial zone that is largely inaccessible to the transit and radial velocity methods, which are heavily biased toward finding planets in tight, close-in orbits.

Perhaps the most exciting and unique capability of microlensing is its ability to detect planets that are not orbiting a star at all. Theories of planet formation predict that some planets may be ejected from their parent systems during chaotic early stages, doomed to wander the galaxy alone as "rogue planets." These worlds are cold, dark, and essentially invisible to all other methods. However, if a rogue planet passes in front of a background star, its gravity can produce a microlensing event all on its own. Since the planet's mass is small, the event will be very short, lasting only a few hours to a couple of days. In September 2020, astronomers announced the detection of OGLE-2016-BLG-1928, a microlensing event with a timescale of just 42 minutes, consistent with a free-floating planet of roughly Earth's mass. Microlensing is currently the only viable technique for finding this hidden population of nomadic worlds.

However, the method is not without its significant drawbacks. The most obvious is that microlensing events are exceedingly rare and completely random. For every million stars monitored, only a handful will undergo a lensing event at any given time. The probability that one of those events will also betray the signature of a planet is lower still. This dependence on chance alignments makes planet detections difficult and unpredictable.

The second major disadvantage is that the events are one-offs. A transit is periodic, and a radial velocity signal repeats with every orbit, allowing for confirmation and refined measurements over time. A microlensing event, however, happens once and never again. This makes follow-up observations of the planet itself impossible. We get a single, fleeting glimpse, and then the cosmic alignment is broken forever.

This leads to a related problem: characterizing the system. During a microlensing event, the light from the foreground lens star is often blended with the much brighter, magnified light of the background source star. The lens star itself is frequently a faint, distant object that is difficult or impossible to observe on its own. Without a good measurement of the lens star's mass and distance, it's difficult to convert the measured planet-to-star mass ratio into a definitive mass for the planet. Astronomers can sometimes wait years for the lens star to drift far enough away from the source star to be observed and characterized by powerful instruments like the Hubble Space Telescope or the Keck Observatory, as was eventually done for the planet OGLE-2005-BLG-169Lb.

To overcome the rarity of these events, microlensing requires a coordinated global effort. Survey teams like OGLE, MOA, and KMTNet act as sentinels, issuing alerts to the astronomical community when a stellar lensing event begins. This allows a network of follow-up telescopes, such as PLANET and MicroFUN, to devote intensive, round-the-clock monitoring to the ongoing event, hoping to catch the short-lived planetary spike if it occurs. This global teamwork is essential to capturing the high-resolution data needed to properly model the planetary anomaly.

The first confirmed planet discovered using this technique was OGLE-2003-BLG-235Lb/MOA-2003-BLG-53Lb, announced in 2004. It was found to be a Jupiter-mass planet orbiting its star from a distance of several AU. Since then, the method has uncovered dozens of worlds, including systems that bear a striking resemblance to our own. The OGLE-2006-BLG-109L system, for instance, was found to host two gas giants, a Jupiter-analogue and a Saturn-analogue, in orbits that mirror the configuration of our own solar system's gas giants.

The future of microlensing is incredibly bright, thanks in large part to a single upcoming mission: NASA's Nancy Grace Roman Space Telescope. Set to launch in the mid-2020s, Roman will conduct a massive microlensing survey from space. Unhindered by Earth's atmosphere or the day/night cycle, Roman will stare at the dense star fields of the galactic bulge for long periods, achieving unprecedented sensitivity and cadence. It is expected to detect more than a thousand exoplanets via microlensing, including many with masses as low as Mars. Roman will complete the census of exoplanets by providing robust statistics on the populations of worlds in wide orbits, giving us our first clear picture of how common planets like Jupiter, Saturn, and even Earth are at Earth-like distances from their stars. It will single-handedly revolutionize our understanding of cold planets and provide a definitive count of the lonely rogue planets wandering our galaxy.

CHAPTER SIX: Capturing the Faint: Direct Imaging of Exoplanets

To an astronomer, there is little that can compare to the thrill of capturing an image. The indirect methods of exoplanet detection, for all their stunning ingenuity and success, leave a certain desire unfulfilled. Finding a shadow or measuring a wobble is a triumph of inference, like identifying an animal by its tracks in the snow. Direct imaging, however, is the scientific equivalent of seeing the creature itself. It is the ambitious, audacious goal of capturing actual photons of light that have bounced off or been emitted by a planet orbiting another star. It is the quest to take a picture of another world.

For decades, this goal remained firmly in the realm of science fiction. The challenge is almost comically daunting and can be distilled into two core problems: contrast and separation. A planet is an almost infinitesimally faint object sitting right next to an incomprehensibly bright one. A Sun-like star outshines an Earth-like planet in reflected visible light by a factor of about ten billion to one. Even a self-luminous gas giant, glowing from the heat of its own formation, is still a million to a billion times fainter than its star.

This immense contrast is compounded by a tiny angular separation. Because the stars are so far away, the angle between a star and its orbiting planet as seen from Earth is minuscule, often smaller than the angle subtended by a coin viewed from hundreds of kilometers away. Trying to resolve this faint speck of planetary light from the blinding glare of its parent star has been famously compared to trying to spot a firefly hovering next to a searchlight from thousands of kilometers away. It is a technical hurdle of the highest order, and overcoming it has required the development of some of the most advanced optical technologies ever created.

The primary tool for battling the star's glare is an instrument with a deceptively simple name: the coronagraph. Invented in the 1930s by French astronomer Bernard Lyot to study the Sun's faint outer atmosphere, the corona, its principle is straightforward. A coronagraph is essentially a small, precisely machined occulting mask placed in the focal plane of a telescope. This mask physically blocks the light from the star, creating an artificial eclipse inside the instrument. With the overwhelming starlight suppressed, the much fainter light from any orbiting companions can, in theory, shine through.

In practice, it is far more complex. Even with a perfect mask, the wave nature of light causes some of the starlight to diffract, or bend, around the edges of the mask, spreading out and contaminating the field of view where a planet might be hiding. Modern coronagraphs are incredibly sophisticated systems, employing a series of masks and stops to manage this diffracted light. Designs with names like the vortex coronagraph and the shaped-pupil coronagraph use advanced optical physics to channel the starlight away from the target area with breathtaking efficiency, aiming to create a "dark hole" of deep contrast right where a planet is expected to be.

For telescopes on the ground, a coronagraph alone is not enough. The Earth's turbulent atmosphere constantly bends and distorts incoming starlight, causing it to twinkle and smear out. This blurring effect, known as "seeing," spreads the star's light over a wide area, making it impossible to focus it neatly onto a tiny coronagraphic mask. The solution to this atmospheric chaos is a technology known as adaptive optics, or AO. An AO system is a marvel of real-time engineering that effectively un-twinkles the stars.

An AO system works by first measuring the incoming atmospheric distortion. It does this by observing a reference star—either the target star itself or a nearby guide star—and analyzing how its light is being warped. This information is fed to a computer hundreds or even thousands of times per second. The computer then sends commands to a "deformable mirror" in the telescope's light path. This mirror, often with hundreds or thousands of tiny actuators on its back, can change its shape with incredible speed and precision, creating a surface that is the exact inverse of the atmospheric distortion. The result is that the distorted light waves bounce off the corrective mirror and emerge perfectly flat, as if they had traveled through the vacuum of space. This allows the star's image to be focused into a tight, sharp point, making it possible for the coronagraph to do its job effectively.

Even with the best coronagraphs and adaptive optics, the image is never perfectly clean. Imperfections in the telescope's own mirrors and optics scatter starlight, creating a complex pattern of residual light called "speckles." This speckle pattern is often brighter than the planet being sought and can easily create false detections. To solve this final piece of the puzzle, astronomers turn to clever observational strategies and sophisticated software algorithms, a field known as high-contrast post-processing.

One of the most powerful of these techniques is Angular Differential Imaging (ADI). In this method, the telescope is programmed to keep the target star perfectly centered, while the orientation of the telescope on the sky is allowed to rotate. From the telescope's point of view, the star remains fixed, and so does the speckle pattern created by the telescope's optics. The planet, however, being a separate object in the sky, appears to rotate around the star in an arc. By taking a long sequence of images and then using software to de-rotate them so the planet's position lines up, the images can be combined. The static speckles, which rotate with the telescope, get smeared out and can be subtracted away, while the planet's signal, which stays in one place in the de-rotated frame, adds up and becomes visible.

A complementary technique is Spectral Differential Imaging (SDI), which takes advantage of the fact that a planet's light has a different color profile, or spectrum, than its star. For example, a cool gas giant planet will have deep absorption bands in its atmosphere from molecules like methane, which are absent in the much hotter star. By taking simultaneous images at a wavelength where methane absorbs strongly and at a nearby wavelength where it doesn't, astronomers can subtract one image from the other. The starlight speckles, which look nearly identical at both wavelengths, cancel out. The planet, however, is bright at one wavelength and faint at the other, so it remains as a clear signal in the subtracted image. In practice, many modern instruments use a combination of these and other advanced techniques to dig a planetary signal out of the noise.

When these technologies work in concert, the scientific payoff is immense. A direct image is more than just a picture; it is a source of rich data that is unobtainable by other means. For starters, it provides unambiguous proof of a planet's existence. Furthermore, by taking images over a period of months and years, astronomers can literally watch the planet move in its orbit. This allows for a direct measurement of the orbit's size, shape, and orientation in space. Unlike the radial velocity method, which suffers from the $\sin i$ ambiguity, direct imaging allows for a clear determination of the orbital inclination, which in turn helps to nail down the planet's true mass.

Perhaps the greatest power of direct imaging lies in what can be done once the planet's light is isolated. That light can be passed through a spectrograph, allowing astronomers to perform detailed characterization of the planet itself. Unlike transmission spectroscopy, which can only probe the thin sliver of atmosphere at the planet's terminator during a transit, direct imaging spectroscopy samples the light from the planet's entire observed hemisphere. This light contains a wealth of information about the planet's physical conditions.

By analyzing the spectrum of a directly imaged planet, scientists can determine its temperature and estimate its overall energy budget. They can identify the chemical fingerprints of various molecules in its atmosphere, detecting the presence of water vapor, methane, carbon monoxide, carbon dioxide, and ammonia. These chemical inventories provide crucial clues about how and where the planet formed. The spectrum can also reveal the presence and properties of clouds or hazes, telling us about the planet's weather. By observing subtle changes in brightness as the planet rotates, it may even be possible to measure the length of its day.

Given the immense challenges, the direct imaging method comes with a strong observational bias. It works best for planets that are easiest to see, which means planets that are very bright and very far from their stars. The planets that best fit this description are young, massive gas giants. They are young because a newly formed planet is still incredibly hot, radiating away the immense gravitational energy it accumulated during its formation. This makes it glow brightly in infrared wavelengths, where the contrast with its parent star is most favorable. They are massive because a bigger planet is simply brighter and easier to see. And they orbit far from their star because the greater separation makes it easier to distinguish their light from the star's glare.

Consequently, direct imaging surveys are finding a completely different class of planet than those found by the transit and radial velocity methods. Instead of hot Jupiters in tight, days-long orbits, direct imaging is revealing a population of "cold Jupiters" and "super-Jupiters," massive gas giants several times the mass of Jupiter, orbiting their stars at distances comparable to or greater than Saturn and Neptune in our own solar system. This technique, therefore, doesn't compete with the other methods; it complements them, filling in a crucial piece of the planetary census by exploring the cold, outer frontiers of other solar systems.

The first object widely accepted as a directly imaged exoplanet was found in 2004, orbiting not a true star but a brown dwarf—a "failed star" not massive enough to sustain nuclear fusion. A team using the Very Large Telescope (VLT) in Chile captured an image of a faint companion to the brown dwarf 2M1207. The companion, named 2M1207b, was determined to be about five times the mass of Jupiter. The discovery was confirmed a year later when observations showed that the faint object was indeed moving through space with the brown dwarf, proving it was a gravitationally bound companion and not a distant background star.

The true watershed moment for the field arrived in 2008. An international team of astronomers using the Keck and Gemini telescopes in Hawaii announced the direct imaging of not one, but three planets orbiting the star HR 8799, a young star about 130 light-years away in the constellation Pegasus. A fourth planet was discovered in the same system two years later. The HR 8799 system was a spectacular revelation. Here was a complete planetary system, with four massive gas giants orbiting their star in wide, stable orbits, like a scaled-up version of our own outer solar system. For the first time, astronomers had a moving picture of another solar system, as subsequent observations over the following decade have beautifully traced the stately orbital dance of these four worlds.

Another landmark system is Beta Pictoris, a very young star surrounded by a vast disk of dust and debris, a planetary system in the making. In 2008, a French team confirmed the presence of a planet, Beta Pictoris b, orbiting within this disk. The planet, about eight times the mass of Jupiter, was found on an orbit roughly the size of Saturn's. Thanks to its nearly edge-on orientation, astronomers were even able to watch the planet pass behind its star in 2017, providing a wealth of data about both the planet and its surrounding environment.

These discoveries have been made possible by a new generation of powerful, specialized instruments mounted on the world's largest ground-based telescopes. These include the Spectro-Polarimetric High-contrast Exoplanet REsearch (SPHERE) instrument on the VLT and the Gemini Planet Imager (GPI) on the Gemini South telescope. These instruments combine advanced adaptive optics, high-performance coronagraphs, and sophisticated detectors into integrated planet-hunting machines, pushing the limits of contrast ever deeper.

The new titan of direct imaging, however, resides in space. The James Webb Space Telescope (JWST), with its large, cold mirror and stable location beyond the Earth's atmosphere, is a formidable tool for this work. While not solely a planet-imager, JWST is equipped with several state-of-the-art coronagraphs. In September 2022, it captured its first direct image of an exoplanet, a gas giant named HIP 65426 b. JWST's unparalleled sensitivity allows it not only to image planets but, more importantly, to perform spectroscopy on them with unprecedented detail across a wide range of infrared wavelengths, opening a new window into the chemistry of their atmospheres.

Looking further into the future, astronomers are dreaming even bigger. One concept for future space telescopes is the starshade, or external occulter. This would involve a separate, large, petal-shaped spacecraft flying in precise formation tens of thousands of kilometers away from a space telescope. This starshade would be positioned to cast a perfect, ultra-dark shadow of the target star over the telescope, blocking the starlight before it even enters the instrument. This technique could potentially achieve the ten-billion-to-one contrast needed to directly image an Earth-like planet in reflected light. Missions based on this concept, like the proposed Habitable Worlds Observatory, represent the ultimate goal of direct imaging: to capture a pale blue dot orbiting a distant star, and to search its light for the telltale signs of a habitable, and perhaps even an inhabited, world.

CHAPTER SEVEN: A Rogue's Gallery: The Variety of Known Exoplanets

For millennia, our concept of a "planet" was crafted from a sample size of eight. We had the small, dense, rocky worlds of the inner solar system and the large, billowy gas and ice giants of the outer regions. It was a neat, orderly, and deeply ingrained pattern, one that astronomers naturally assumed would be the standard blueprint for solar systems everywhere. The discovery of thousands of worlds orbiting other stars has torn that blueprint to shreds. The sheer, unbridled variety of planets in our galaxy has proven to be more bizarre, more extreme, and more wonderful than we could have imagined. We have discovered that our solar system is not the rule, but merely one possible outcome of planetary formation. The galactic census has revealed a veritable rogue's gallery of worlds, forcing us to rethink our neat categories and expand our definitions of what a planet can be.

The main classes of exoplanets are broadly defined by their size and mass. At the top of the scale are the **Gas Giants**, planets composed primarily of hydrogen and helium, similar to Jupiter and Saturn. The first major surprise of the exoplanet era was that many of these giants were found in orbits that were shockingly close to their stars. These are the famous **Hot Jupiters**, gas giants that orbit their stars in as little as a few days, placing them nearer to their suns than Mercury is to our own. The first exoplanet found around a Sun-like star, 51 Pegasi b, is the archetypal example, a massive world with a four-day year. Roasted to extreme temperatures, their atmospheres are thought to feature exotic clouds of molten rock and stellar compositions more akin to cool stars than anything in our solar system.

Not all gas giants are so close to their parent star. We are also finding **Cold Jupiters**, massive planets that orbit in the colder, outer regions of their systems, in paths more analogous to our own Jupiter. These worlds are much harder to detect with the transit and radial velocity methods due to their long orbital periods, but they are beginning to be revealed by direct imaging and gravitational microlensing, painting a more complete picture of these large planetary systems.

The next step down in size brings us to the **Neptune-like planets**. As the name suggests, these are worlds similar in size to Uranus or Neptune, likely possessing a core of rock and metal surrounded by a thick envelope of hydrogen and helium. Just as with the gas giants, astronomers have found "Hot Neptunes" that orbit very close to their star, and a particularly common class of planet that is entirely absent from our solar system: the **Mini-Neptune**. These worlds are larger than Earth but smaller than Neptune, representing a transitional phase between rocky and gaseous planets. Their composition is a subject of intense study; they may be "gas dwarfs" with rocky cores and puffy hydrogen atmospheres, or they could be water worlds with deep global oceans.

Perhaps the most common type of planet in the galaxy is one our solar system completely lacks: the **Super-Earth**. This category refers to planets with a mass higher than Earth's but substantially below that of Neptune. The term refers only to a planet's mass and size, not to any Earth-like surface conditions or habitability. Super-Earths are a truly diverse bunch. Depending on their formation and location, they could be dense, rocky planets, magma-covered lava worlds, or possess thick, crushing atmospheres that make them more like small gas planets than large terrestrial ones. The sheer abundance of this class of planet is a profound cosmic puzzle, raising the question of why our own solar system is missing what appears to be the galaxy's most popular model.

Finally, we have the **Terrestrial** planets, worlds that are Earth-sized or smaller and are primarily composed of rock and metal. These are the planets that often capture the most public attention, as they represent the most likely candidates in the search for life. Finding them is a monumental challenge due to their small size and mass, but missions like Kepler have shown they are common. The famous TRAPPIST-1 system, for example, features seven Earth-sized rocky planets orbiting a small, cool star. Several of these worlds orbit within the star's habitable zone, the region where liquid water could potentially exist on the surface. While far from a guarantee of life, the discovery of such systems demonstrates that worlds with a similar scale to our own are not a rarity in the cosmos.

Beyond these broad classifications lies a collection of worlds so strange they seem plucked from the pages of science fiction. These are the true rogues, the planets that defy easy categorization and highlight the extreme outcomes possible in the universe. Among the most dramatic are the **Lava Worlds**. These are typically rocky planets, often super-Earths, that orbit so perilously close to their star that their surfaces are permanent oceans of molten magma. A prime example is 55 Cancri e, a super-Earth with a surface temperature hot enough to melt rock, which completes an orbit in a mere 18 hours.

Theorists also predict the existence of **Carbon Planets**, sometimes dubbed "diamond planets." These are thought to form in protoplanetary disks that are rich in carbon rather than oxygen. Instead of the silicate rocks that make up Earth, their geology would be dominated by carbides and graphite. Under sufficient pressure, a thick layer of diamond could form deep beneath the surface. While no definitive carbon planet has been confirmed, a few candidates exist, including the pulsar planet PSR J1719-1438 b, which may be the remnant core of a star that has been transformed into a diamond world.

On the other end of the density spectrum are the bizarre **Puffy Planets**, or "Super-Puffs." These are planets with masses only a few times that of Earth, yet they have radii as large as Jupiter's, giving them a mean density comparable to that of styrofoam or cotton candy. The planets of the Kepler-51 system are the best examples, enormous in size but incredibly low in mass. How these feather-light worlds can even exist is a major puzzle. One theory suggests they may be surrounded by vast, dusty outflows or perhaps even enormous ring systems that make them appear much larger than they actually are.

Just as some planets are surprisingly puffy, others are the dense, stripped-down remnants of once-mighty worlds. These are the hypothetical **Chthonian Planets**, named after the deities of the Greek underworld. A chthonian planet is thought to be the exposed rocky or metallic core of a gas giant that orbited too close to its star. Over millions of years, the star's intense radiation and stellar winds would have blasted away the planet's thick hydrogen and helium atmosphere, leaving only the dense heart behind. The exoplanet CoRoT-7b, a scorching-hot world orbiting extremely close to its star, has been suggested as a possible candidate for this class of planetary ghost.

The family portraits of planetary systems have also been redrawn. We have discovered **Circumbinary Planets**, which orbit not one, but two stars at once. The planet Kepler-16b was the first definitive "Tatooine" world to be discovered, a cold gas giant that would offer its hypothetical inhabitants a double sunset. Several such worlds have now been found, confirming that stable planetary orbits are possible even in the complex gravitational environment of a binary star system.

Finally, we must not forget the planets that started it all. The very first exoplanets confirmed were discovered in 1992, not around a stable, Sun-like star, but around a pulsar—the rapidly spinning, hyper-dense corpse of a star that has gone supernova. The planets of the PSR B1257+12 system are small, rocky worlds that likely formed from the debris disk left behind after their star's violent death. These **Pulsar Planets** remain some of the rarest and most extreme worlds known, born from ashes in an environment bathed in intense radiation. Their existence was an early and powerful lesson that planets can form in the most unexpected of places.

CHAPTER EIGHT: Hot Jupiters and Super-Earths: A Closer Look at Exotic Worlds

The initial flood of exoplanet discoveries did more than just add new dots to our cosmic map; it shattered the very map itself. The neat, familiar architecture of our own solar system was revealed to be just one of many possible designs, and perhaps not even the most common one. Among the thousands of new worlds, two classes in particular stood out, not just for their exotic nature, but because their very existence posed a fundamental challenge to our understanding of how planets are born and evolve. These were the Hot Jupiters and the Super-Earths. One was a type of planet we thought we understood, found in a place we thought was impossible. The other was a class of planet we never even knew existed, yet it appears to be the most common type of world in the galaxy. Taking a closer look at these two revolutionary discoveries reveals the central puzzles that have driven exoplanet science for decades.

The first of these cosmic enigmas to be discovered was the Hot Jupiter. The name is blunt but perfectly descriptive: these are planets with a mass similar to or greater than our own Jupiter, but they orbit their parent stars in a blistering, intimate embrace. While Jupiter takes nearly twelve years to circle our Sun, a typical Hot Jupiter completes its year in less than ten days, and some do it in under twenty-four hours. The discovery of 51 Pegasi b in 1995, a Jupiter-mass world with a 4.2-day orbit, was the shot that started the revolution. It was not just a new planet; it was a gauntlet thrown down to the entire field of planetary science.

The problem was not that a gas giant could exist, but where it was found. The established theory of planet formation, known as the core accretion model, was built around the evidence from our solar system. This model requires the existence of a "frost line" or "snow line" in the young protoplanetary disk of gas and dust surrounding a newborn star. Inside this line, closer to the star, it is too warm for volatile compounds like water, ammonia, and methane to freeze into solid ice grains. Outside the frost line, in the colder regions, these ices are abundant. It is these icy grains that are believed to be the crucial ingredient for building giant planets.

The theory holds that in the outer disk, these sticky ice particles quickly clump together, forming large planetary cores of perhaps ten Earth masses. Once a core reaches this critical mass, its gravity becomes powerful enough to rapidly pull in the vast amounts of hydrogen and helium gas surrounding it, creating a gas giant. Inside the frost line, however, there is only rock and metal, which are much less abundant. The building blocks are smaller and less sticky, so while rocky planets can form, they never grow massive enough quickly enough to trigger the runaway gas accretion needed to become a Jupiter. According to everything we thought we knew, a gas giant had no business being anywhere near its star.

Yet, there 51 Pegasi b was, and hundreds more have been found since. Their existence demanded an explanation, and it came in the form of a new and transformative concept: planetary migration. Planets, it turns out, do not necessarily stay put in the orbits where they are born. The protoplanetary disk that gives them life is not a serene environment; it's a dynamic, turbulent place, and the gravitational interactions within it can cause a planet's orbit to shrink or grow dramatically over millions of years.

Two main theories have emerged to explain how a Jupiter can end up in a stellar furnace. The first, and more sedate of the two, is called disk migration. In this scenario, the planet forms peacefully beyond the frost line, just as the models predict. However, while it is still embedded in the gaseous disk, it creates powerful spiral waves in the surrounding material. These waves of gas exert a gravitational torque back on the planet, causing it to lose orbital energy and spiral slowly inward toward the star. This process continues until the gas disk dissipates, or until some other mechanism halts the migration, parking the planet in its new, tight orbit. It's a bit like a ship dragging its anchor along the seabed, slowly grinding to a halt close to shore.

The second theory is far more chaotic and violent. Known as high-eccentricity migration, it suggests that Hot Jupiters are the survivors of a planetary game of billiards. In this model, a gas giant forms in the outer system, often along with other massive planets. Over time, their mutual gravitational nudges destabilize their orbits. This can lead to a period of intense chaos where planets are scattered into new, highly elliptical paths, with some being ejected from the system entirely. A planet that gets scattered into a comet-like orbit that brings it extremely close to the star at its point of closest approach will experience powerful tidal forces. The star's gravity will stretch and deform the planet, dissipating orbital energy and causing the orbit to become less elliptical and more circular over time, but at a much smaller radius. This "tidal circularization" effectively reels the planet in, leaving it stranded in a close, hot, circular orbit.

Evidence for both mechanisms exists. Many Hot Jupiters are found to have orbits that are misaligned with the spin of their star, and some even orbit backward. Such skewed orbits are difficult to explain with the gentle inward spiral of disk migration but are a natural consequence of the violent scattering and tilting involved in high-eccentricity migration. On the other hand, the discovery of many systems with multiple, tightly packed planets in neat, circular orbits seems to favor a less chaotic migration process. It is likely that both pathways contribute to the population of Hot Jupiters we see today.

Whatever their journey, the destinations these planets arrive at are truly hellish environments. With surface temperatures soaring above 1,000 Kelvin, and sometimes exceeding 2,500 Kelvin—hotter than the surface of some stars—their physics are unlike anything in our solar system. The intense heat causes their atmospheres to puff up, making many Hot Jupiters significantly larger in radius than our own Jupiter, even if they have less mass. These are some of the lowest-density planets known, giant, bloated spheres of scorching gas.

Being so close to their star, Hot Jupiters are almost certainly tidally locked, with one hemisphere in perpetual, baking daylight and the other in eternal night. This creates an extreme temperature gradient that drives ferocious winds, possibly moving at supersonic speeds, whipping from the day side to the night side. These winds are thought to carry heat around the planet, preventing the night side from freezing and keeping it merely "very hot" instead of "incandescent." This global heat redistribution is a major area of study, with telescopes like the James Webb Space Telescope measuring the temperature map of these worlds to trace the flow of their alien weather systems.

The chemistry of these atmospheres is equally exotic. The day sides are so hot that molecules we think of as forming rocks can exist as vapor. Astronomers have found the spectral signatures of silicates in the atmospheres of some Hot Jupiters, suggesting the presence of clouds made of tiny droplets of molten rock—sand, essentially. On slightly cooler Hot Jupiters, there could be clouds of iron or corundum. Imagine a world where the haze is made of vaporized ruby and sapphire, and where, on the cooler night side, these minerals could condense and fall as rain.

This proximity to the star also means Hot Jupiters are being constantly blasted by intense radiation and stellar wind. This stellar onslaught can be powerful enough to strip away the planet's atmosphere over time, a process known as photoevaporation. Astronomers have observed some Hot Jupiters, like the famous HD 209458 b, trailing vast comet-like tails of escaping hydrogen. This process may be so efficient that over billions of years, a gas giant could be stripped of its entire gaseous envelope, leaving behind only its dense core. This could be the origin of the so-called Chthonian planets, the scorched, airless remnants of worlds that were once mighty gas giants.

If the Hot Jupiter was the planet we knew in a place we didn't, the Super-Earth was the planet we never even knew existed. Defined as worlds with a mass greater than Earth's but less than Neptune's, they are the middleweights of the cosmos, occupying a gap in our own solar system's lineup. Before the exoplanet discoveries, we had no inkling that such a class of planet was not only possible but, as data from the Kepler mission has overwhelmingly shown, likely the most common type of planet in the entire Milky Way. Their sheer abundance is a staggering revelation. For every star like our Sun, there is a good chance—perhaps one in three—that it hosts at least one Super-Earth. Their absence from our own planetary family makes our solar system look increasingly peculiar.

The central mystery surrounding Super-Earths is their composition. What are they made of? A planet's mass alone is not enough to answer this question. The crucial second piece of information is its radius, which can be measured if the planet transits its star. With both mass and radius in hand, astronomers can calculate a planet's bulk density, our best clue to its internal makeup. When the known Super-Earths are plotted on a mass-radius diagram, they don't all fall on a single line. Instead, they fan out, revealing a fascinating diversity of worlds.

Some Super-Earths are quite dense. These are likely to be true rocky worlds, scaled-up versions of Earth and Venus. They would have iron cores and silicate mantles, but with some profound differences. A larger mass means stronger surface gravity. It could also power more vigorous and long-lasting geological activity, like volcanism and plate tectonics, which are thought to be important for regulating a planet's climate and cycling nutrients. A more massive rocky world might also be much better at holding on to a thick atmosphere.

On the other end of the spectrum are the less-dense Super-Earths. These low-density worlds are one of the biggest puzzles in planetary science and could represent several different kinds of planets. Some may be "water worlds," planets that are composed of a significant fraction of water, perhaps up to 50% by mass. Such a world would have a rocky core, but it would be buried beneath a global ocean hundreds or even thousands of kilometers deep. The pressure at the bottom of such an ocean would be immense, great enough to crush liquid water into exotic forms of high-pressure ice, like Ice VII or Ice X, which are solid even at very high temperatures.

Alternatively, these low-density Super-Earths may not be watery at all. They could instead be "gas dwarfs," or Mini-Neptunes. These would be planets with a solid, Earth-like core that, during its formation, managed to attract and hold onto a thick, puffy envelope of hydrogen and helium gas. This gaseous blanket would make the planet much larger in radius for its mass, giving it a low overall density. The surface of such a world would not be solid; it would be a blurry transition into a deep, crushing atmosphere, with pressures and temperatures far too high for life as we know it.

Distinguishing between a water world and a gas dwarf is incredibly difficult, as both can produce similar mass and radius measurements. The answer likely lies in analyzing their atmospheres. The James Webb Space Telescope is now capable of taking this next step, sniffing the air of these worlds to see if it is dominated by water vapor, hinting at a steamy ocean world, or by hydrogen and helium, the tell-tale sign of a Mini-Neptune. The initial findings seem to suggest that many of these low-density worlds do indeed have extended, hydrogen-rich atmospheres.

Unlike Hot Jupiters, which are often found as lonely companions to their stars, Super-Earths are frequently found in bustling, compact multi-planet systems. The iconic TRAPPIST-1 system is a prime example, with seven rocky, Earth-sized planets (some of which fall into the Super-Earth mass range) orbiting their star in orbits smaller than Mercury's. These systems are like planetary peas in a pod, often exhibiting a remarkable regularity in the spacing of their orbits, a phenomenon known as orbital resonance. This suggests they likely formed together and migrated inward gently, a stark contrast to the chaotic scattering that may have formed many Hot Jupiters.

The discovery that the galaxy's most common planets may be Super-Earths has profound implications for the search for life. On one hand, a rocky Super-Earth could be a "super-habitable" world. Its stronger gravity could help it retain a thicker atmosphere, offering better protection from cosmic rays. Its greater internal heat could drive plate tectonics for billions of years longer than on Earth, helping to maintain a stable climate. On the other hand, this same high gravity could be a curse, leading to a crushing atmosphere that creates a runaway greenhouse effect, or it could prevent the formation of continents altogether, resulting in a deep water world with a potentially sterile seabed. The line between a Super-Earth and a Mini-Neptune is a critical one, likely representing the boundary between a potentially habitable rock and an uninhabitable gas world. Understanding this dividing line is now one of the foremost goals in exoplanet science.

CHAPTER NINE: The Building Blocks: Planet Formation and Evolution

The stunning diversity of planets now known to populate the galaxy, from scorching Hot Jupiters to the teeming systems of Super-Earths, all spring from remarkably humble beginnings. Every planetary system, our own included, is born from the leftovers of star formation. The process begins inside vast, cold, dark clouds of interstellar gas and dust. When a dense pocket within one of these molecular clouds collapses under its own gravity, it forms a spinning protostar at its center. Due to the conservation of angular momentum—the same principle that makes a spinning ice skater speed up when they pull their arms in—not all of the collapsing material falls directly onto the new star. Instead, a significant fraction flattens out into a vast, rotating disc of gas and dust that surrounds the stellar newborn. This is a protoplanetary disk, the crucible in which planets are forged.

These disks are immense, often stretching for hundreds of astronomical units, but they are also incredibly tenuous. Their composition largely mirrors that of the original molecular cloud: about 99% of their mass is gas, primarily hydrogen and helium, with the remaining 1% consisting of tiny solid grains of dust. These dust particles are not like household dust bunnies; they are microscopic motes of silicates, carbon compounds, and iron, similar in consistency to smoke or soot. In the cold outer regions of the disk, beyond a boundary known as the "frost line" or "snow line," temperatures are low enough for volatile compounds like water, methane, and ammonia to freeze into solid ice crystals, dramatically increasing the amount of solid material available. It is from this scant one percent of solid matter that the entire menagerie of planetary worlds must be built.

The first step in this grand construction project is to get the microscopic dust grains to stick together. In the early, relatively gentle environment of the disk, this happens through simple processes. Tiny grains bump into each other and cling together due to electrostatic forces, much like how a balloon rubbed on your hair will stick to a wall. Through countless collisions over thousands of years, these fluffy aggregates gradually grow from the size of smoke particles to pebbles, and then to larger objects. This initial phase of growth, however, runs headlong into a significant problem that for decades puzzled planetary scientists: the "meter-size barrier".

This barrier is actually a collection of several obstacles. As an object grows to around a meter in size, its interaction with the surrounding gas of the disk changes. The gas in the disk is partially supported by its own pressure, causing it to orbit the star slightly slower than a solid object would at the same distance. This means a meter-sized boulder is constantly plowing through a gaseous headwind, which causes it to lose orbital energy and spiral rapidly inward toward the central star. The effect is strongest for objects of this size, and the timescale is terrifyingly short; a meter-sized body at Earth's orbit would fall into its star in just a few hundred to a thousand years, far too quickly to grow into a planet. Furthermore, collisions between these fast-moving boulders are more likely to be destructive, shattering them back into smaller fragments rather than building them up.

For planets to exist, there must be a way to leapfrog this treacherous stage. One of the leading solutions to this puzzle is a mechanism known as the streaming instability. This theory suggests that the interaction between the solid particles and the gas is more complex than a simple headwind. The drag felt by the solids can cause them to clump together into dense filaments and swarms. Within these traffic jams, the concentration of solid particles can become so high that their collective gravity takes over, causing them to collapse directly into much larger bodies, from tens to hundreds of kilometers in diameter. This process essentially allows the building blocks to skip the dangerous meter-size phase entirely, jumping straight from pebbles to veritable mountain-sized objects called planetesimals. These planetesimals are the true building blocks of planets.

Once planetesimals have formed, the second act of planet formation begins, dominated by gravity. The largest of these planetesimals, the "winners" of the initial scramble, begin to gravitationally influence their neighbors. Their weak but persistent pull draws in smaller planetesimals, leading to a period of runaway growth. The bigger an object gets, the stronger its gravity becomes, and the faster it can sweep up material in its orbital path. In this way, kilometer-sized planetesimals grow into moon-sized or Mars-sized planetary embryos, also known as protoplanets. From this point, the path to a finished planet can diverge, following one of two main theoretical routes: core accretion or gravitational instability.

The core accretion model is the most widely accepted theory for how the majority of planets, including all the planets in our solar system, are formed. It is a "bottom-up" process. It posits that a planetary embryo continues to grow by accreting nearby planetesimals and pebbles. In the inner part of the protoplanetary disk, where only rock and metal are solid, this process builds up terrestrial planets like Earth and Venus. The growth is relatively slow because solid material is scarce. An embryo's growth is complete when it has swept its orbital path clean of other building blocks.

In the outer disk, beyond the frost line, the story is different. The abundance of ice means that planetary embryos can grow much larger and faster. The core accretion model predicts that once a solid core reaches a critical mass, typically around 5 to 10 times the mass of the Earth, a dramatic new phase begins. At this mass, the core's gravity is strong enough to start pulling in not just solids, but also the huge amounts of hydrogen and helium gas that dominate the disk. This triggers a runaway process of gas accretion. The more gas the planet pulls in, the more massive it becomes, and the more effectively it pulls in even more gas. In a geological blink of an eye, the solid core shrouds itself in a vast, thick atmosphere, becoming a gas giant like Jupiter or an ice giant like Neptune.

The second, more dramatic theory of planet formation is known as the gravitational instability model. This is a "top-down" approach that may explain the formation of very massive gas giants found at great distances from their stars. In this scenario, planets form directly from the gas of the disk, bypassing the slow, bottom-up construction of a solid core. The theory proposes that in the early stages, if a protoplanetary disk is sufficiently massive and cold, its own self-gravity can cause it to become unstable and fragment. Dense, spiral arms can form within the disk, and if a region within one of these arms becomes dense enough, it can collapse under its own gravity, rapidly forming a planet-sized clump of gas in a few thousand years—a timescale much faster than core accretion.

Whether these clumps survive to become stable planets is a matter of intense debate. For the collapse to proceed, the clump must be able to cool and radiate away its heat efficiently; otherwise, internal pressure will push it apart. The gravitational instability model is thought to be a less common mode of planet formation than core accretion, but it provides a plausible pathway for the creation of the massive super-Jupiters that have been directly imaged in the cold, distant reaches of other solar systems, where the core accretion process would likely be too slow.

The story of a planet is not over once it has formed. The protoplanetary disks in which they are born are dynamic, evolving environments, and planets are rarely stationary. The discovery of Hot Jupiters, massive planets in scorching orbits where they could not possibly have formed, provided the first definitive proof that planets must move from their birthplaces. This process is called planetary migration. There are several mechanisms that can drive this cosmic relocation.

For planets still embedded in the gas disk, their own gravity creates waves in the disk material, similar to the wake of a boat moving through water. The gravitational pull of these waves, in turn, tugs on the planet, causing it to exchange angular momentum with the disk. For lower-mass planets like Super-Earths, this interaction (called Type I migration) generally causes them to spiral inward. For massive, Jupiter-sized planets that are powerful enough to open a gap in the disk, the planet becomes locked to the gas and gets dragged inward as the disk itself slowly loses energy and accretes onto the star. This slower process is known as Type II migration.

Migration doesn't have to be a smooth and gentle process. After the gas disk has dissipated, planetary systems can enter a period of violent instability. The gravitational interactions between the newly formed planets can destabilize their orbits, leading to a chaotic free-for-all. In this game of planetary billiards, planets can be thrown into highly eccentric, comet-like orbits, or be ejected from the system entirely to become rogue planets. A planet scattered into an orbit that brings it very close to its star can then have its orbit circularized by powerful tidal forces, stranding it as a Hot Jupiter. This violent "planet-planet scattering" is thought to be responsible for the many eccentric orbits we observe among exoplanets.

The final stage of system formation involves clearing out the last of the debris. Over a period of a few million years, the primordial gas that dominated the protoplanetary disk is eventually removed. This happens through a combination of factors. Some of the gas is accreted by the planets or the central star. The rest is blown away by the young star's increasingly powerful radiation and stellar winds, a process called photoevaporation. Once the gas is gone, planet formation largely ceases.

What remains is a young, often unstable planetary system and a disk of leftover planetesimals, similar to our own asteroid and Kuiper belts. This can lead to a final, tumultuous period of impacts. In our own solar system, the migration of the giant planets is thought to have triggered an event known as the Late Heavy Bombardment, where a huge number of asteroids and comets were flung into the inner solar system, scarring the surfaces of the Moon, Mercury, and Mars. This rain of fire, while destructive, was also a delivery service, potentially bringing water and organic compounds to the young terrestrial planets.

The complex interplay of these processes—the initial conditions of the disk, the competition between different formation models, the type and extent of migration, and the chaos of late-stage impacts—is what ultimately sculpts the final architecture of a planetary system. It explains why we see such a bewildering variety of outcomes: some stars with lonely, scorching Hot Jupiters; others with orderly, compact systems of Super-Earths in neat resonant orbits; and still others with widely-spaced gas giants like our own. Each exoplanet system is a fossil record of these dynamic and often violent building processes, a unique solution to the universal challenge of building worlds from dust and gas.

CHAPTER TEN: Inside and Out: Understanding Exoplanet Interiors and Surfaces

The discovery of a new exoplanet is a moment of profound inference. Across light-years of empty space, astronomers capture a whisper of a world—a fleeting shadow, a gravitational tug, a brief flash of magnified light. From these faint signals, they can deduce a planet's orbit, estimate its mass, and measure its size. But what of the planet itself? What lies beneath its alien sky? Is its surface solid rock, a global ocean, or a swirling vortex of gas with no surface at all? We cannot send a probe to drill into its crust or a rover to traverse its plains. Every insight we have into the physical nature of these distant worlds is a masterpiece of scientific deduction, a journey from a handful of photons to a portrait of a planet's heart.

The first and most fundamental key to unlocking a planet's inner secrets is its bulk density. This single value, derived from the powerful combination of the transit and radial velocity methods, is our initial guide to a planet's composition. The transit of a planet across its star reveals its physical radius, telling us how large it is. The gravitational wobble it induces in that same star allows us to measure its mass, telling us how much "stuff" it contains. By simply dividing the mass by the volume (calculated from the radius), we arrive at the planet's average density. This is the moment a data point transforms into a tangible world.

A planet's density immediately places it into a broad category. A very low density, less than that of water, strongly suggests a gas giant, a world like Jupiter or Saturn composed primarily of lightweight hydrogen and helium. A high density, similar to that of Earth (which is 5.5 grams per cubic centimeter), points toward a rocky, terrestrial world. An intermediate density might hint at an "ice giant" like Neptune or, more exotically, a world with a substantial fraction of its mass locked up in water. This first clue is plotted on a mass-radius diagram, a foundational chart in exoplanet science. On this graph, astronomers plot theoretical lines showing where planets of a pure composition—pure iron, pure rock, pure water, pure hydrogen—would lie. By seeing where a newly discovered planet falls on this plot, we can make our first educated guess about what it is made of.

Of course, a planet is not a uniform ball of a single substance. Bulk density is just an average, masking a complex and layered interior. The real challenge, and the focus of immense computational effort, is to build physically plausible models of a planet's internal structure that match its known mass and radius. Scientists become planetary architects, using the laws of physics as their blueprints. They begin with principles like hydrostatic equilibrium—the balance between the inward crush of gravity and the outward push of internal pressure—and combine them with "equations of state," complex formulas that describe how materials like iron, silicate rock, water, and hydrogen behave under the extreme pressures and temperatures found deep inside a planet.

The process is one of trial and error. A modeler might start with a core of a certain size and composition (say, 30% iron) and wrap it in a mantle of a different composition (70% silicate rock). They then run the model to see if a stable planet with that structure results in the correct total mass and radius. The problem is that there is often more than one way to get the right answer. This is known as the degeneracy problem. For example, a low-density Super-Earth could be a large rocky planet with a small iron core, or it could be a smaller rocky planet with a massive iron core that is hidden beneath a deep, low-density ocean of water. Differentiating between these possibilities is a major frontier in exoplanet characterization.

For gas giants, the models paint a picture of truly extreme environments. At their heart likely lies a core of rock, metal, and ice, although its exact nature is a subject of intense debate. This core might not be solid and well-defined like Earth's. Under the immense pressure and temperature, the boundary between the core and the envelope above may be a "fuzzy," partially dissolved region where rock and hydrogen mix. Surrounding this core is the planet's main constituent: hydrogen. In the outer layers, it exists as a gas, but as you descend deeper, the pressure becomes so colossal—millions of times Earth's atmospheric pressure—that it transforms hydrogen into a bizarre and exotic state of matter known as metallic hydrogen. In this state, the electrons are stripped from the hydrogen atoms and can flow freely, causing the hydrogen to behave like a liquid metal. The powerful magnetic fields of planets like Jupiter are thought to be generated by electrical currents flowing within this vast, planet-sized ocean of metallic hydrogen.

Hot Jupiters present an additional puzzle. Many of them have radii that are significantly larger than predicted by standard models for a planet of their mass. They appear "puffy" or "inflated." The leading explanation is the intense heat they receive from their parent star. This constant, powerful irradiation is thought to pump energy deep into the planet's interior, heating it up and causing its atmosphere to expand like a loaf of bread in a hot oven. Other mechanisms, such as tidal heating from slightly elliptical orbits or the slowing of the planet's cooling by atmospheric opacity, may also contribute to keeping these giants bloated.

The interiors of Neptune-like worlds and the enigmatic Mini-Neptunes are thought to be different again. These planets are often called "ice giants," but the term can be misleading. Their structure is likely a three-layer cake: a solid rock-and-metal core at the center, a deep hydrogen and helium atmosphere on the outside, and a thick, soupy mantle of "ices" in between. These are not ices in the familiar sense of a frozen cube. They are a hot, dense, fluid mixture of water, methane, and ammonia, compressed to such a degree that they behave very differently from their counterparts on Earth. It is within this hot, slushy mantle that some of the universe's strangest chemistry may take place, and it represents the key difference between a gas giant and an ice giant.

For the rocky Super-Earths, modeling their interiors is an exercise in pushing geology to its limits. A more massive Earth-like planet would have far greater pressures at its core and in its mantle.

These pressures can force the minerals in the rock to undergo phase transitions, changing their crystal structure to become more compact. For instance, the mineral perovskite, common in Earth's lower mantle, would itself be compressed into an even denser form called post-perovskite under the weight of a Super-Earth's mantle. These changes in mineralogy can have profound effects on how the planet's interior behaves, influencing everything from the way heat flows out of the core to whether the planet can sustain long-term geological activity like plate tectonics.

Understanding a planet's interior is a quest for the unseen, but these models provide crucial context for what we might find on the surface. For gas giants and many Mini-Neptunes, the very concept of a surface is meaningless. There is no solid ground to stand on, only a seamless transition where the gaseous atmosphere grows ever denser and hotter with depth, eventually becoming a supercritical fluid where the distinction between gas and liquid dissolves. For these worlds, there is no "out," only "in."

For other planets, however, the surface is a real place, a boundary between geology and atmosphere, and we are beginning to map its features from afar. The most extreme examples are the lava worlds. These are rocky planets, like 55 Cancri e and CoRoT-7b, orbiting so close to their star that their dayside temperatures are high enough to melt silicate rock. Their surfaces are likely vast, churning oceans of magma. We can infer this not only from their orbital distance but also by observing their thermal phase curve. As the planet orbits its star, we see its different faces—its hot day side, its cooler terminator (the line between day and night), and its night side. By measuring the total infrared light from the system, astronomers can map the planet's heat. For many lava worlds, these maps show an incredibly hot spot on the side permanently facing the star, consistent with a molten surface that radiates heat inefficiently to the night side.

Then there are the potential water worlds, planets whose low density suggests they are composed of a significant fraction of water. A hypothetical world with 50% water by mass would be unlike anything in our solar system. It would be covered in a single, global ocean thousands of kilometers deep. The pressure at the bottom of such an ocean would be staggering, millions of pounds per square inch. Under this immense pressure, the water at the bottom would be forced into exotic forms of high-pressure ice, such as Ice VII or Ice X. These ices are solid even at hundreds of degrees Celsius. This would create a bizarre geology where the "seafloor" is not rock, but a thick, solid layer of ice separating the liquid ocean above from the silicate mantle below, a scenario with profound implications for the kind of chemical exchange between rock and water that may have been crucial for the origin of life on Earth.

For terrestrial planets in more temperate orbits, like the famous worlds of the TRAPPIST-1 system, discerning the nature of their surfaces is the ultimate challenge. We can infer that they are solid and rocky, but what are they like? A key factor is tidal locking. Because these planets orbit so close to their small, cool stars, they are almost certainly tidally locked, with one hemisphere in perpetual daylight and the other in perpetual night. This creates an "eyeball" climate state. The surface facing the star could be a baked desert, while the far side is a frozen wasteland of ice and darkness. The most interesting real estate on such a world might be the terminator, the twilight ring between day and night, where temperatures could be mild enough for liquid water to exist.

A planet's surface and interior are also constantly shaped by its relationship with its star and its sibling planets. An exoplanet in a slightly eccentric orbit is continuously flexed by its star's changing gravitational pull. This tidal flexing generates immense frictional heat in its interior, a process known as tidal heating. This can power intense and long-lasting volcanism, as seen on Jupiter's moon Io, constantly resurfacing the planet and venting gases that can form and replenish an atmosphere. In this way, a planet's orbit can directly control its geological activity.

Conversely, the star can un-make a surface. The intense X-ray and ultraviolet radiation from a star can heat a planet's upper atmosphere and blow it away into space, a process called photoevaporation. Observations have revealed a curious gap in the population of known planets, an "evaporation valley." There are plenty of Super-Earths and plenty of Mini-Neptunes, but very few planets in between. This suggests that stellar radiation may be a powerful sculptor of planets. A world that starts as a Mini-Neptune, with a rocky core and a thick hydrogen envelope, could have its entire atmosphere stripped away over billions of years, leaving behind only the naked, airless core—a planet transformed from a gas dwarf into a barren Super-Earth. The journey inward, from sparse data to a rich understanding of planetary guts and geophysics, is a testament to how much we can learn from so little, painting pictures of alien worlds from the inside out.

CHAPTER ELEVEN: Alien Atmospheres: Characterizing the Air of Other Worlds

The solid body of a planet is its foundation, a massive and mysterious realm of crushed rock and exotic metals that lies forever beyond our direct view. Its atmosphere, however, is its face, turned outward to the cosmos. This thin, often tenuous envelope of gas is the most accessible part of an exoplanet, the dynamic interface between the world and its star. It is the atmosphere that dictates the climate on the surface, that shields the ground from harsh radiation, and that traps the heat needed to maintain liquid water. It is the realm of clouds and winds, of weather and chemistry. Most importantly, it is in the air of other worlds that we expect to find the most telling clues about a planet's nature, its history, and its potential to harbor life.

Before we can even begin to study an alien atmosphere, we first need to know that it exists. Not every planet has one. A small, rocky world orbiting too close to its star or one that is not massive enough to gravitationally bind its gases might be a barren, airless rock like our own Moon or Mercury. The first clue to the presence of an atmosphere often comes from a planet's bulk density, a fundamental property derived from its mass and radius. A planet with a very high density is likely a solid sphere of rock and metal with little to no gaseous envelope. But a planet with a low density, especially a very low density, is a dead giveaway. To be so large for its mass, it must be puffed up by a substantial atmosphere. The "Super-Puff" planets of the Kepler-51 system are an extreme example, worlds as large as Jupiter but with only a few times the mass of Earth, making their average density comparable to that of cotton candy. Such objects cannot be solid; they must be dominated by vast, extended atmospheres.

The origin of a planet's atmosphere is intimately tied to its formation and evolution, and generally follows one of two paths. The first is the path of the giants. Gas giants and ice giants, which form in the cold outer regions of a protoplanetary disk, grow massive enough to gravitationally capture huge amounts of gas directly from the nebula around them. They therefore possess "primary" atmospheres, primordial envelopes composed almost entirely of the most abundant elements in the universe: hydrogen and helium. These atmospheres are fossils of the disk from which the entire system was born. Many Mini-Neptunes, the class of planet between Earth and Neptune, are also thought to have held onto these primordial hydrogen blankets, making them essentially scaled-down gas dwarfs.

Terrestrial planets, being too small and too close to their star to hold onto this lightweight gas, follow a different path. They form as largely airless bodies of rock and metal. Their atmospheres are "secondary," meaning they are acquired long after the planet has formed. This can happen in two main ways. The first is through volcanic outgassing. As the planet's interior churns and geological activity commences, trapped gases like carbon dioxide, water vapor, nitrogen, and sulfur compounds are vented from volcanoes, gradually building up an atmosphere from the inside out. The second method is delivery from space. During the chaotic early history of a solar system, the inner planets are bombarded by comets and asteroids, which are rich in water and other volatile compounds. This celestial rain can deliver the essential ingredients for an atmosphere and oceans. Earth's own atmosphere is a classic example of a secondary atmosphere, dominated by nitrogen that likely seeped out from its interior over geological time.

The diversity of these formation pathways leads to a spectacular variety of atmospheric types, each with its own unique chemistry and physics. The atmospheres of Hot Jupiters are laboratories for extreme chemistry. Their primary composition is hydrogen and helium, but they are cooked to temperatures exceeding a thousand, or even two thousand, degrees Celsius. At these temperatures, chemistry becomes bizarre. Instead of water clouds, astronomers expect to find clouds made of rock-forming minerals. On the "cooler" Hot Jupiters, these may be silicate clouds—essentially clouds of vaporized sand. On hotter worlds, they could be clouds of iron or exotic minerals like corundum, the substance that makes up rubies and sapphires. Finding the spectral fingerprints of these substances allows astronomers to take the temperature of these worlds with remarkable precision.

These infernal worlds are also almost certainly tidally locked, with one side in perpetual day and the other in perpetual night. The immense temperature difference between the baked day side and the merely scorching night side drives the most extreme winds in the known universe. Atmospheric models predict that heat is ferried from day to night by winds that could reach speeds of several kilometers per second, far exceeding the speed of sound. This global circulation prevents the night side from cooling off completely and creates strange temperature maps, which we can now begin to observe. Rather than the hottest point being directly under the star, these powerful winds often shift the hot spot eastward, a clear sign of a dynamic, churning atmosphere.

For the abundant Super-Earths and Mini-Neptunes, the nature of their atmospheres is one of the most pressing questions in exoplanet science. It represents a fundamental dividing line. A low-mass planet that has retained a thick primary atmosphere of hydrogen is a gas dwarf, or Mini-Neptune. It would have no solid surface, just a deep, crushing atmosphere where pressures and temperatures quickly become inhospitable. On the other hand, a planet in this same mass range that has lost its primary atmosphere and developed a secondary one could be a true Super-Earth. It might be a water world with a thick, steamy atmosphere of water vapor, or a terrestrial world with an atmosphere dominated by carbon dioxide or nitrogen. Determining which of these scenarios is more common is crucial for understanding how many potentially habitable worlds are out there. The line between a rocky planet and a gassy one appears to be one of the most important boundaries in nature.

The ultimate goal, for many, is to characterize the atmospheres of Earth-sized planets orbiting within their star's habitable zone. Here, the search is for an atmosphere that feels familiar, one that could support a climate suitable for liquid water. This means looking for a secondary atmosphere, likely dominated by an inert gas like nitrogen, which provides bulk and pressure. The key ingredients we search for within this bulk atmosphere are greenhouse gases like carbon dioxide and, most importantly, water vapor. The presence of water vapor is a critical first check for habitability, as it is the source of liquid water for oceans, lakes, and rivers. But an atmosphere is more than just a list of ingredients; its structure is just as important.

The structure of an atmosphere is defined by its temperature and pressure profile—how these properties change with altitude. On Earth, our atmosphere is divided into layers. We live in the troposphere, the lowest layer, where weather happens and temperature decreases with height. Above that is the stratosphere, where temperature begins to rise with altitude due to the absorption of ultraviolet radiation by the ozone layer. This temperature inversion is a key feature of our atmosphere. Finding a similar thermal inversion in the atmosphere of an exoplanet could be a vital clue. It would tell us that there is some gas in its upper atmosphere that is very good at absorbing light from its star. While on Earth that gas is ozone, on other worlds it could be something different, but its presence would signal complex chemistry at play.

Of course, alien atmospheres are not just clear gas. Like our own sky, they are filled with clouds and hazes, which can be both a nuisance and a source of valuable information. Clouds form when a particular gas cools and condenses into liquid droplets or solid ice particles. On Earth, this is water. On a scorching Hot Jupiter, it might be iron. On a frigid gas giant far from its star, it could be methane. The presence and altitude of cloud decks tell us about the temperature at different layers of the atmosphere. Hazes are a different beast, formed when starlight triggers chemical reactions in the upper atmosphere, creating a smog of complex organic molecules. The hazy, orange sky of Saturn's moon Titan is a prime example in our own solar system.

For astronomers, clouds and hazes are a double-edged sword. A thick, high-altitude layer of clouds or haze can completely obscure the gases below, making it impossible to probe the atmosphere's deep composition. Some planets appear to have "flat" spectra, a sign that our view is blocked by such a layer. However, the composition of the cloud particles themselves can be revealing. Furthermore, by observing how the brightness and color of a planet change over time, we can get a sense of its cloud cover. We might see the periodic brightening of a particularly reflective patch of clouds as the planet rotates, allowing us to map its weather patterns and even measure its rotation period—the length of its day.

The primary method for studying these distant atmospheres is to let the planet's star do the work for us. As a transiting planet passes in front of its star, a tiny fraction of the starlight shines through the planet's atmospheric limb. As this light passes through the gas, molecules in the atmosphere absorb very specific colors, or wavelengths, of that light. Each molecule, whether it's water, methane, or sodium, has a unique spectral "barcode" of colors it absorbs. By measuring which colors of starlight are missing after the transit, astronomers can identify the chemical ingredients of the alien air. This powerful technique, called transmission spectroscopy, allows us to perform a chemical inventory from light-years away. One of the very first atmospheric detections using this method was the discovery of sodium in the air of the Hot Jupiter HD 209458 b in 2002.

A complementary technique uses the planet's own light. Just before a planet disappears behind its star in a secondary eclipse, or occultation, our telescopes receive the combined light of the star and the planet's day side. When the planet is hidden, we receive light from only the star. By subtracting the in-eclipse light from the out-of-eclipse light, we can isolate the light that was coming directly from the planet. This is the planet's thermal emission, its incandescent glow. Analyzing the spectrum of this emitted light reveals the temperature of the planet's day side and can also show the spectral fingerprints of the molecules in its atmosphere, providing a crucial check on the results from transmission spectroscopy.

For a smaller number of planets, those that are very large and very far from their stars, it is possible to capture their light directly. As discussed previously, direct imaging separates the planet's light from the star's glare, allowing for detailed spectroscopic analysis. This is particularly powerful because it allows us to study the atmospheres of planets that do not transit. The James Webb Space Telescope has revolutionized all three of these methods, providing stunningly detailed spectra and giving us our first real glimpse into the chemistry of these distant worlds. It has successfully measured the atmospheric composition of gas giants, Mini-Neptunes, and even begun to probe the air of rocky Super-Earths, searching for the molecules that could signal a temperate and potentially habitable world.

Finally, by watching the total light from a system continuously throughout a planet's entire orbit, we can construct a "phase curve." As the planet orbits, we see its different faces, from its fully illuminated day side to its dark night side, just like the phases of our Moon. The changing brightness tells a story. We can map the temperature distribution across the planet, pinpointing its hottest and coldest regions. These maps are our first crude weather reports for other worlds, revealing how efficiently their atmospheres circulate heat and providing the raw data needed to test our complex models of alien climates. From a simple point of light, we are beginning to paint a picture of skies on worlds we will never visit, tasting their air from across an ocean of space.

CHAPTER TWELVE: Reading the Rainbow: Spectroscopic Analysis of Exoplanet Atmospheres

Of all the tools in the astronomer's arsenal, none is more powerful or more versatile than spectroscopy. It is the art and science of decoding the messages hidden within light. We are accustomed to thinking of light in terms of its brightness, but its true richness is revealed when it is broken apart into its constituent colors, a process that happens naturally when sunlight passes through a prism or raindrops to create a rainbow. This spread of colors, from the deepest red to the most intense violet, is a spectrum. For an astronomer, a spectrum is not just a pretty pattern; it is a treasure trove of data, a detailed chemical and physical report sent across light-years of empty space. When applied to the faint light trickling from the atmospheres of exoplanets, spectroscopy becomes a form of cosmic alchemy, transforming tiny flickers of data into a detailed inventory of alien air.

At its heart, the technique relies on a fundamental discovery made in the 19th century: different chemical elements, when heated, emit or absorb light only at specific, characteristic wavelengths, or colors. This creates a unique spectral "barcode" for every atom and molecule. The bright, continuous rainbow of light emitted by the hot, dense core of a star is known as a continuous spectrum. When that light passes through the cooler, more tenuous gas of the star's own outer layers, atoms in that gas absorb their specific colors, creating a pattern of dark lines superimposed on the rainbow. This is an absorption spectrum, and it is precisely what allows us to know that our Sun is made of hydrogen, helium, and a smattering of heavier elements.

Exoplanet spectroscopy applies this exact same principle, but with a crucial twist. It uses the entire star as a backlight. The planet's atmosphere becomes the "cool gas," and by studying how it filters the starlight passing through it, we can read its composition. The science behind these spectral barcodes is rooted in the strange rules of quantum mechanics. Within an atom, electrons cannot orbit the nucleus at just any distance; they are confined to specific energy levels, like rungs on a ladder. To jump from a lower rung to a higher one, an electron must absorb a photon of light with the exact amount of energy needed to make that leap. Because a photon's energy is directly tied to its wavelength, this means an atom can only absorb very specific colors of light. When we see a dark line in a spectrum, we are seeing the shadow of atoms all performing the same quantum leap.

Molecules add wonderful layers of complexity to this picture. In addition to having electronic energy levels like single atoms, a molecule—a collection of atoms bound together—can also vibrate and rotate. The chemical bonds between atoms can stretch, compress, and bend, and the entire molecule can spin like a top. Each of these motions is also "quantized," meaning they can only happen at specific, discrete energy levels. The energy jumps between these vibrational and rotational states are much smaller than the jumps between electron shells, and they correspond to photons in the infrared part of the spectrum. This is why molecules like water, methane, and carbon dioxide have rich, complex absorption spectra in the infrared, with whole bands made of thousands of individual lines, creating a dense and unique fingerprint. It is this molecular complexity in the infrared that makes the James Webb Space Telescope, which is optimized for these wavelengths, such a revolutionary tool for atmospheric characterization.

The most prolific method for reading these fingerprints is transmission spectroscopy, which can only be used for the lucky fraction of planets that transit their stars. The technique is conceptually straightforward but technically demanding. Astronomers use a spectrograph to carefully measure the spectrum of the host star just before the planet begins its transit. This gives them a baseline measurement of the star's pure, unfiltered light. Then, as the planet passes in front of the star, they take another spectrum. During the transit, a tiny fraction of the starlight shines through the upper layers of the planet's atmosphere along its edge, or terminator. The atoms and molecules in this atmospheric ring absorb their characteristic wavelengths, carving their dark absorption lines into the starlight.

By dividing the in-transit spectrum by the out-of-transit spectrum, astronomers can cancel out the star's own spectral features, isolating the faint absorption signature of the planet's atmosphere alone. The result is a transmission spectrum, a graph that shows how much light is blocked at each wavelength. Where the graph peaks, the atmosphere is most opaque. In essence, the planet appears slightly larger at the wavelengths where its atmosphere is absorbing light most strongly. By mapping these peaks, scientists can identify the substances present in the alien air. The first great success of this method came in 2002 with the detection of the element sodium in the atmosphere of the Hot Jupiter HD 209458 b, a landmark achievement that proved it was possible to read the chemistry of another world.

Of course, the universe is rarely so simple. The signal from transmission spectroscopy is incredibly tiny, often just a few parts per million of the total starlight. Teasing this faint signal from the noise is a formidable challenge. A major source of confusion is the star itself. Stars are not perfect, uniform spheres of light. They are covered in dark, cool starspots and bright, hot patches called faculae. As the star rotates, these features can move across its face. If a transiting planet happens to pass over a starspot, it is blocking a region that is cooler and dimmer than the rest of the star. This can alter the overall color of the starlight in a way that can mimic or contaminate the absorption signal from the planet's atmosphere. Disentangling the planet's signal from the star's own activity requires meticulous monitoring and sophisticated modeling.

A powerful complementary method gets around the need for a backlight by capturing the planet's own glow. This is emission spectroscopy, performed during a planet's secondary eclipse, or occultation, when it disappears behind its star. Just before the planet vanishes, our telescopes receive the combined light from both the star and the planet's hot, glowing day side. While the planet is hidden, we see only the light from the star. By subtracting the "in-eclipse" measurement from the "out-of-eclipse" measurement, astronomers can isolate the light that was coming solely from the planet itself.

This light tells a different, but equally important, story. It is a thermal emission spectrum, revealing the temperature and composition of the planet's day side. By fitting a theoretical blackbody radiation curve to the spectrum, astronomers can directly measure the planet's temperature. The spectral features observed are also different. While transmission spectroscopy reveals absorption lines from the cool terminator, emission spectroscopy can show either absorption or emission lines from the hot day side. If the atmosphere is cooler at the top and hotter at the bottom, like Earth's troposphere, we see absorption lines as light from the deeper, hotter layers passes through the cooler upper layers.

However, some Hot Jupiters have revealed a fascinating twist: a temperature inversion, or stratosphere. In these atmospheres, some chemical species in the upper layers absorb so much incoming starlight that they heat the top of the atmosphere to a higher temperature than the layers below. This is analogous to Earth's ozone layer. In such a case, these hot upper-atmosphere molecules will glow brightly at their characteristic wavelengths, creating bright emission lines on top of the planet's thermal spectrum. The detection of these inversions, often linked to the presence of light-absorbing molecules like titanium oxide and vanadium oxide, provides a direct probe of the atmosphere's thermal structure and energy balance.

While transit and eclipse spectroscopy have been the workhorses of the field, they are not the only tools. A more advanced technique, known as high-resolution spectroscopy, trades breadth for detail. Instead of observing a wide range of colors with low resolution, it zooms in on a very narrow range of wavelengths, measuring the precise shape and position of individual spectral lines with exquisite detail. This precision allows astronomers to leverage the Doppler effect to great advantage. As a planet orbits its star, its motion causes its spectral lines to be Doppler-shifted—

blueshifted as it moves toward us, and redshifted as it moves away. The star's spectral lines, by contrast, remain relatively stationary.

By searching for a faint set of spectral lines that shifts back and forth with the known orbital period of the planet, astronomers can detect the chemical fingerprint of its atmosphere, even if the planet doesn't transit. Furthermore, the shape of the spectral line itself contains information about the planet's dynamics. The rotation of the planet will broaden the line, potentially allowing for a measurement of the length of its day. Moreover, powerful winds carrying gas from the planet's hot day side to its cooler night side will create an additional Doppler signature, with one side of the planet appearing slightly blueshifted and the other slightly redshifted. By carefully modeling this effect, astronomers have been able to make the first direct measurements of wind speeds in the atmospheres of other worlds.

The ultimate prize is direct imaging spectroscopy. When astronomers manage the incredible feat of separating a planet's light from its star's glare, that precious trickle of photons can be directed into a spectrograph. This provides a spectrum of the light reflected or emitted from the planet's entire visible hemisphere, rather than just the thin sliver of the terminator probed by transmission spectroscopy. This gives a more global and potentially more representative view of the planet's atmospheric composition. It is particularly powerful for studying the chemical inventories of young, self-luminous gas giants in wide orbits. The ratio of key elements in their atmospheres, such as the ratio of carbon to oxygen, can serve as a chemical fossil, providing clues about where in the original protoplanetary disk the planet formed relative to the snow lines of different chemical species.

No matter which technique is used, the final spectrum is rarely a clean, unambiguous message. The data from telescopes like Hubble and JWST is complex, noisy, and often degenerate, meaning different combinations of atmospheric properties could produce very similar spectra. To interpret this data, scientists rely on a powerful computational process known as atmospheric retrieval. This is essentially a process of sophisticated trial and error. A computer is programmed to generate millions of hypothetical planetary atmospheres, each with a different temperature profile, a different mix of chemical abundances, and different properties for clouds and hazes. For each of these millions of models, the computer calculates what its spectrum would look like.

Then, using robust statistical frameworks, such as Bayesian inference, the computer compares all of these theoretical spectra to the actual observed data. It then determines which set of atmospheric parameters provides the best fit. The result is not a single, definitive answer, but rather a set of probabilities for each property. For instance, a retrieval might conclude that the abundance of water vapor is most likely between 100 and 500 parts per million, and that a cloud deck is very likely to be present below a certain altitude. It is through this rigorous, model-driven process that the faint light from another world is translated into a detailed weather report, providing our first glimpse into the incredible diversity of climates and chemistries that populate the skies of the galaxy.

CHAPTER THIRTEEN: The Goldilocks Zone: The Search for Habitable Worlds

In the grand cosmic quest to find life beyond Earth, one substance has, above all others, guided our search: liquid water. It is not necessarily that life everywhere *must* be based on water, but it is the only basis for life we know. Water is a superb solvent, capable of dissolving and transporting the chemical ingredients necessary for biology, and it remains liquid across a relatively wide range of temperatures. For this reason, the search for other worlds that might harbor life has largely become a search for worlds that could host liquid water on their surfaces. This seemingly simple requirement has given rise to one of the most powerful and evocative concepts in modern astronomy: the habitable zone.

It is a term with a storybook charm, often called the "Goldilocks Zone." The name captures the concept perfectly. It is the orbital region around a star where a planet with a suitable atmosphere could maintain a surface temperature that is not too hot and not too cold, but just right for liquid water to exist. A planet orbiting too close to its star would be a scorching furnace, its oceans boiled away into space, a scenario reminiscent of Venus. A planet orbiting too far from its star would be a deep freeze, its water locked up in a global shell of ice, a fate similar to that of Mars. The habitable zone is that temperate, circumstellar sweet spot where a world might have lakes, rivers, and oceans.

The location and size of this zone are dictated almost entirely by the star itself. A star is a celestial furnace, and the amount of heat a planet receives depends on how hot that furnace is and how close the planet huddles to it. A bright, massive star pours out enormous amounts of energy, pushing its habitable zone far outward. For a star like this, a planet in an Earth-like orbit would be vaporized; its Goldilocks region might lie at a distance comparable to Jupiter's orbit in our own system. Conversely, a small, dim star radiates only a feeble amount of energy. Its habitable zone is a narrow, intimate ring, nestled extremely close to the star, far inside the orbit of Mercury.

For a star like our Sun, a G-type main-sequence star, the habitable zone is a comfortably wide band. Current estimates place its inner edge somewhere just beyond the orbit of Venus and its outer edge somewhere near the orbit of Mars. Earth, of course, sits comfortably within these boundaries. The Sun's relative stability is also a key factor. It has been shining with a reasonably consistent output for billions of years, providing a long and stable period during which the complex chemistry of life could potentially arise and evolve. For this reason, sun-like stars were long considered the prime targets in the search for another Earth.

As our surveys have expanded, however, we have found that nature has other preferences. The most common type of star in the Milky Way, by a huge margin, is not the sun-like G-type, but the much smaller, cooler, and dimmer M-type star, commonly known as a red dwarf. These stars make up perhaps seventy-five percent of the galaxy's stellar population. If you were to pick a star at random, chances are it would be a red dwarf. This statistical reality forces us to consider the habitability of the planets that orbit them, and it presents a fascinating suite of problems.

Because red dwarfs are so faint, their habitable zones are located incredibly close to the star. A planet in the Goldilocks zone of a typical red dwarf might have a year that lasts only a few weeks or even just a few days. This extreme proximity creates a host of potential challenges for life. The first and most significant is tidal locking. Just as the Earth's gravity has locked one face of the Moon permanently toward us, the intense gravity of a nearby red dwarf would almost certainly lock its habitable-zone planets into a synchronous rotation. This would result in one hemisphere being trapped in perpetual, searing daylight and the other in unending, frozen night.

Such a planet would be a world of two extremes, often called an "eyeball planet." The side facing the star could be a baked desert, while the dark side would be so cold that the planet's atmosphere could freeze out and collapse onto the surface as ice. Life, if it existed, might be confined to the terminator, the twilight ring between the two hemispheres, where temperatures could be more clement. It is possible that a sufficiently thick atmosphere could circulate enough heat from the day side to the night side to prevent this atmospheric collapse and create a more globally mild climate, but this remains a major point of debate among climate modelers.

The second major hurdle for life around red dwarfs is their violent tempers. These small stars, especially in their youth, are known for their ferocious stellar flares. They can suddenly and unpredictably erupt, blasting their nearby planets with torrents of high-energy X-rays and charged particles. A single powerful flare could be strong enough to strip away a planet's atmosphere over time and bathe its surface in sterilizing radiation, making it extraordinarily difficult for life to get a foothold, let alone survive. Any life on such a world would need powerful protection, perhaps living deep underground or in oceans shielded from the deadly stellar tantrums.

Even the quality of light from a red dwarf is different. Unlike the Sun, which emits a broad spectrum of light, a red dwarf's output is heavily skewed toward the red and infrared end of the spectrum. Photosynthesis on Earth has evolved to be most efficient using the blue and red light our Sun provides. Life on a red dwarf planet would need to evolve a different kind of biological machinery, perhaps using pigments that are black in appearance in order to absorb as much energy as possible across a wider range of wavelengths. A forest on such a world might not be green, but a deep, dark shade of purple or black.

The classical definition of the habitable zone, based purely on stellar heat, is a wonderfully useful guide, but it is also a dramatic oversimplification. It treats the planet as a passive, inert billiard ball being warmed by a star. In reality, the planet itself is an active and crucial participant in setting its own surface temperature. The most important planetary factor is its atmosphere. An atmosphere acts as a thermal blanket, trapping heat through the greenhouse effect. A planet with a thick atmosphere rich in greenhouse gases like carbon dioxide or methane can stay warm and wet even at the distant, chilly outer edge of the habitable zone.

Our own planet's history provides the perfect case study. Billions of years ago, the Sun was significantly fainter and cooler than it is today—perhaps only seventy-five percent as bright. According to the simplest models, the young Earth should have been a frozen snowball. Yet, geological evidence clearly shows that liquid water and life were present. This puzzle, known as the "Faint Young Sun Paradox," is resolved by considering Earth's early atmosphere. It was likely much thicker and far richer in greenhouse gases like methane and carbon dioxide, providing a much stronger greenhouse effect that compensated for the dimmer Sun. This demonstrates that a planet's climate is not just a matter of location, but of its specific atmospheric and geological properties.

A planet's size also plays a critical role. A more massive rocky planet—a Super-Earth—has a stronger gravitational pull, which helps it to hold on to its atmosphere against the stripping forces of stellar winds and flares. Its greater mass also means it would have a hotter interior for a longer period of time. This could power vigorous and long-lasting geological activity, such as volcanism. Volcanoes are not just destructive forces; they are also crucial for replenishing an atmosphere by venting gases from the planet's interior. This volcanic recycling of materials, part of a process known as the carbonate-silicate cycle on Earth, is thought to be a key mechanism for regulating a planet's climate over geological timescales. A larger planet may simply be better at creating and maintaining the conditions for habitability.

Another critical factor is a planet's albedo, or its reflectivity. A planet covered in dark oceans and rock will absorb more sunlight and be warmer than a planet covered in bright, reflective ice and clouds. Clouds themselves present a complex feedback loop. They reflect incoming sunlight, which has a cooling effect, but they also trap infrared heat radiating from the surface, which has a warming effect. The net result depends on the type, altitude, and coverage of the clouds. These planetary characteristics mean that two planets in the exact same orbit around the exact same star could have wildly different climates.

Despite these complexities, the habitable zone remains our primary tool for sifting through the thousands of known exoplanets to find the most promising candidates for follow-up study. And the discoveries have been thrilling. The famous TRAPPIST-1 system, for instance, is a red dwarf orbited by seven Earth-sized rocky planets. At least three of these—TRAPPIST-1e, f, and g—orbit within the star's habitable zone. They are all tidally locked, but their existence provides an incredible natural laboratory for studying the potential habitability of such worlds. Another landmark is Kepler-186f, the first Earth-sized planet to be discovered in the habitable zone of another star, in this case a red dwarf more than five hundred light-years away. It orbits near the cool outer edge of its zone, earning it the nickname of "Earth's cousin."

Our closest exoplanetary neighbor, Proxima Centauri b, also orbits within the habitable zone of its red dwarf star. The discovery of a potentially temperate world right on our cosmic doorstep was a stunning revelation. However, Proxima Centauri is a known flare star, highlighting the harsh reality that being in the right place is no guarantee of clement conditions. These worlds are not confirmed to be habitable; they are simply confirmed to be in the region where the right kind of planet *could* be habitable. They are the prime targets for the next generation of telescopes, which will attempt the monumental task of scanning their atmospheres for the chemical hints of water, and perhaps even life.

Finally, astronomers have added further layers of nuance to the concept by considering habitability across time and space. The "Continuously Habitable Zone" is a more stringent definition, referring to the region around a star that remains habitable for a long stretch of geological time, perhaps billions of years. Stars brighten as they age, causing their habitable zones to slowly migrate outward. A planet that is in the zone today might have been too hot in the past, or vice-versa. The Continuously Habitable

Zone is the orbital sweet spot that stays "just right" for long enough to allow for the slow process of biological evolution.

On an even grander scale, some have proposed a "Galactic Habitable Zone." This theory suggests that some regions of the Milky Way itself are more conducive to life than others. The galactic center is a violent place, rife with sterilizing radiation from supernovae and the supermassive black hole at its heart. The outer fringes of the galaxy, on the other hand, are poor in heavy elements, the metals and rocks needed to form terrestrial planets in the first place. The Galactic Habitable Zone is the proposed ring in between these extremes—a quiet, element-rich suburb where life-bearing planets might be most likely to form and thrive. It is a reminder that a planet's habitability is not just a local issue, but is tied to the grand, evolving story of its star and its entire galaxy.

CHAPTER FOURTEEN: Beyond the Zone: Habitability in Extreme Environments

The concept of the habitable zone, that comfortable orbital band where a planet is not too hot and not too cold, has served as a primary beacon in our search for life-bearing worlds. It is an intuitive and powerful guide, a cosmic treasure map where 'X' marks the spot for liquid water. For decades, it has focused our attention on planets that look something like Earth, orbiting stars that look something like the Sun. Yet, our own solar system offers tantalizing hints that this map may be incomplete. It suggests that the conditions for life might not be confined to worlds with blue skies and temperate climates, but could flourish in places of perpetual darkness and unimaginable pressure, environments so extreme they force us to redraw the boundaries of biology itself. The galaxy may have habitable real estate far beyond the sunny, temperate suburbs of the Goldilocks zone.

Our first clue to this hidden potential comes from the icy moons orbiting the gas giants in our own backyard. Worlds like Jupiter's moon Europa and Saturn's moon Enceladus orbit far outside the Sun's classical habitable zone. Their surfaces are frigid wastelands, encased in shells of water ice as hard as granite, with temperatures plummeting to hundreds of degrees below zero. By all traditional measures, they should be inert, frozen, and dead. Yet, we have compelling evidence that beneath these forbidding crusts of ice lie vast, dark oceans of liquid water. These are not mere puddles; Europa's subsurface ocean is estimated to contain more than twice the amount of liquid water found in all of Earth's oceans combined. The secret to their warmth is not the distant Sun, but the immense gravitational power of their parent planets.

This phenomenon, known as tidal heating, is a slow and relentless source of energy. As a moon like Europa travels in its slightly elliptical orbit, the gravitational pull from Jupiter is constantly changing, being stronger on the side closer to the planet and weaker on the far side. This differential pull stretches and flexes the entire moon. Imagine bending a metal paperclip back and forth; it quickly becomes warm due to internal friction. The same principle applies to Europa. The constant gravitational kneading generates immense friction and heat deep within its rocky core and icy mantle. This internal furnace, powered by gravity, is more than sufficient to melt the ice from below, creating and maintaining a globe-spanning ocean of liquid water, completely decoupled from the sunlight that warms Earth.

The implications for exoplanetary systems are profound. It means that any gas giant, whether it is a Hot Jupiter or a Cold Jupiter, could potentially host a family of habitable moons. An "Exo-Europa" orbiting a gas giant does not need to be in the star's habitable zone; it only needs to be in the right gravitational relationship with its planet to generate enough tidal heat. This opens up an entirely new parameter space for habitability. We could have life-bearing worlds orbiting stars that are too hot, too cold, or too violent, because the moon's habitability is provided by its planet, not its star. These moons would be worlds without a sunrise, their oceans shrouded in permanent darkness under kilometers of ice, but they would be warm, wet, and potentially teeming with life.

Of course, life in such an environment would be radically different from most life on Earth. Without sunlight, photosynthesis is impossible. The base of the food chain could not be plants or algae. Instead, life would have to rely on chemosynthesis, a process of deriving energy from chemical reactions rather than light. On Earth, we have a perfect analogue for this in the ecosystems surrounding deep-sea hydrothermal vents. These fissures in the ocean floor spew out superheated, mineral-rich water from the planet's interior. In the crushing blackness of the abyss, specialized bacteria thrive on the chemical soup of sulfur compounds erupting from these vents. They form the base of a vibrant and complex food web of tube worms, crabs, and other creatures that never see the Sun. A similar scenario is entirely plausible for a tidally heated ocean world. The interaction between the hot, rocky seafloor and the cold, liquid water could provide a steady supply of the chemical energy needed to sustain a biosphere, a self-contained ecosystem powered by the dance of gravity.

The search for these potentially habitable exomoons is one of the most exciting and challenging frontiers in astronomy. Moons are much smaller and harder to detect than their parent planets, and we are only just beginning to develop the techniques to find them. One potential method is to look for subtle anomalies in a planet's transit light curve. A large moon orbiting a transiting planet would cause a tiny, secondary dip in the starlight or slightly alter the timing and duration of the main transit. Another spectacular possibility is the detection of water vapor plumes. On Saturn's moon Enceladus, great geysers of water vapor and ice crystals erupt from cracks in its southern polar region, venting material from its subsurface ocean directly into space. If an exomoon were doing the same, a telescope like the James Webb Space Telescope could potentially detect the spectral signature of water in the planet's vicinity, a smoking gun for a hidden ocean.

Habitability might not even require the presence of a planet-moon system. The internal heat of a planet itself could be enough. While tidal forces are a powerful engine, planets also have their own heat sources, leftover energy from their violent formation and, more importantly, the long, slow decay of radioactive elements like uranium, thorium, and potassium in their rocky mantles and cores. This geothermal heating is what drives plate tectonics and volcanism on Earth. For a sufficiently massive planet, this internal heat could be substantial enough to maintain liquid water deep underground, even if the surface is frozen solid. This gives rise to the concept of a "lithospheric habitable zone"—a region not of orbital space, but deep within a planet's crust where temperatures are right for water to be liquid.

This idea becomes particularly compelling when we consider the loneliest worlds in the galaxy: rogue planets. These are planets that have been ejected from their parent systems and now wander the cold, dark void of interstellar space, untethered to any star. Their surfaces would be unimaginably cold, approaching absolute zero. Yet, a large terrestrial or super-Earth rogue planet could retain enough geothermal heat for billions of years to maintain a subsurface ocean. Life in such a place would be the ultimate form of isolation, dwelling in a dark, warm ocean, completely oblivious to the starry cosmos above its impenetrable icy roof. Detecting such a world would be incredibly difficult, but the fact that they could exist dramatically expands the galactic footprint for potential life.

Another way a planet could stay warm far from its star is by wearing a very thick blanket. The insulating properties of an atmosphere are critical for retaining heat, and it turns out that some gases are far better insulators than others. The most effective of all is molecular hydrogen. While hydrogen is too light for a small planet like Earth to hold onto, a more massive super-Earth could retain a dense, hydrogen-dominated atmosphere for billions of years. Hydrogen is a potent greenhouse gas, exceptionally good at trapping heat. A planet with such an atmosphere could maintain surface temperatures suitable for liquid water even if it is receiving very little energy from a distant star, or even if it is a rogue planet living off its own internal heat.

This has led to the development of the "Hycean" planet concept, a portmanteau of "hydrogen" and "ocean." These are hypothetical worlds with liquid water oceans under hydrogen-rich atmospheres. The physics of these atmospheres means that the surface pressure would be high, perhaps hundreds of times that of Earth, and the "ocean" might be better described as a supercritical fluid. Nonetheless, the conditions could fall within the bounds of what we consider habitable. Because hydrogen is so light, these atmospheres would be incredibly extended and puffy, making them comparatively easy to detect and characterize with transmission spectroscopy. This is a thrilling prospect, as it suggests that some of the first atmospheres we study in detail could be on these exotic, far-flung worlds, potentially expanding the habitable zone for a given star by a factor of ten.

Finally, we must confront the most speculative but perhaps most important question: must life be based on liquid water at all? Our search is guided by what we know, and all life on Earth uses water as its solvent. But could there be other possibilities? The most famous case study for this is Saturn's largest moon, Titan. Titan is a world of breathtaking strangeness. It is larger than the planet Mercury and possesses a thick, nitrogen-rich atmosphere, denser than Earth's. But it is far too cold for liquid water. At its surface temperature of minus 179 degrees Celsius, water is a rock-hard mineral. Instead, Titan's hydrological cycle is based on hydrocarbons. It has clouds, rain, rivers, lakes, and seas made not of water, but of liquid methane and ethane.

Could some form of exotic, "weird life" exist in these frigid hydrocarbon seas? The chemical challenges are immense. Methane is not as good a solvent as water, and the extreme cold would make metabolic reactions incredibly slow. There would be no oxygen to breathe. Yet, life is fundamentally a process of harnessing chemical energy, and a world like Titan has energy gradients and complex organic chemistry in abundance. We cannot rule out the possibility that life, in a form we can barely imagine, could arise in such an environment. Acknowledging this possibility is a crucial act of scientific humility. It forces us to accept that the Goldilocks zone is a reflection of our own biology. Nature's definition of "just right" might be far broader and more imaginative than we currently dare to assume. The universe could be filled with habitable niches, from the sun-drenched surface of an Earth-like world to the black, crushing depths of a tidally heated moon, to the bizarre methane rivers of an ice-cold titan. Our search has just begun.

CHAPTER FIFTEEN: Water Worlds and Desert Planets: The Potential for Liquid Water

The search for habitable worlds has long been guided by a simple mantra: follow the water. This has led us to the Goldilocks Zone, that orbital region where a planet receives just the right amount of starlight to keep its surface from freezing or boiling. Finding a planet in this temperate zone is a moment of profound excitement, a crucial first step. Yet, location is only half the story. The planet itself holds the final vote on its own habitability, and one of its most important deciding factors is its water budget. How much of this life-giving substance does it actually possess?

The processes of planet formation are messy and chaotic, and there is no reason to assume that every rocky world in a habitable zone ends up with an Earth-like endowment of oceans. A planet's final water content is a lottery, depending entirely on the specific history of its formation and migration. It might accrete a huge fraction of its mass from water-rich materials, or it might form from drier building blocks and receive only a scant delivery of water from comets and asteroids. This fundamental variable leads to a fascinating spectrum of possibilities, bookended by two tantalizing and diametrically opposed archetypes: the water world and the desert planet. One is a world drowning in the very ingredient it needs for life; the other is a world where that same ingredient is a rare and precious resource.

A water world is precisely what its name implies: a planet whose surface is entirely covered by a single, globe-spanning ocean. These are not planets with large seas and a few scattered continents; these are true ocean planets, with no exposed land whatsoever. The idea springs naturally from our understanding of planet formation. In the outer regions of a protoplanetary disk, beyond the frost line, water ice is an abundant building block. A planetary embryo that forms there can easily accumulate a massive inventory of water, potentially making up a significant fraction, perhaps ten percent or more, of its total mass. For comparison, water makes up a minuscule 0.02% of Earth's total mass.

If such a water-rich planet then migrates inward to settle in its star's habitable zone, the ice on its surface would melt, shrouding the planet in a deep, global ocean. The sheer depth of such an ocean would be staggering, far exceeding anything on Earth. While the average depth of Earth's oceans is about four kilometers, the ocean on a water world could be tens or even hundreds of kilometers deep. An astronaut in a submersible descending into such an abyss would be traveling downward for days, passing through a crushing darkness that no glimmer of starlight could ever reach.

The physics of such a deep ocean would be truly alien. The pressure at the bottom would be immense, thousands or even millions of times the atmospheric pressure at sea level on Earth. Under this colossal weight, water ceases to behave like the familiar liquid we know. It is compressed into exotic forms of high-pressure ice. This is not the ice of a frozen winter pond; it is a solid phase of water that is stable even at hundreds of degrees Celsius. A thick layer of this high-pressure ice, with names like Ice VII and Ice X, would form at the base of the ocean, creating a solid "seafloor" of ice that separates the liquid water above from the silicate rocky mantle below.

This scenario presents a profound paradox for habitability. On one hand, the planet has an abundance of the single most important ingredient for life as we know it. On the other, the very structure of the world could stifle the chemistry needed to get life started. On Earth, the continuous interaction between the oceans and exposed continents is thought to be crucial for maintaining a stable, life-friendly climate. The weathering of rocks on land washes essential minerals and nutrients, like phosphorus, into the oceans, providing the raw materials for biology. This process is a key part of the carbonate-silicate cycle, a geological thermostat that regulates Earth's carbon dioxide levels and, therefore, its temperature over millions of years.

On a water world with a high-pressure ice layer at the bottom of its ocean, this vital link between rock and water is severed. The liquid ocean is cut off from the rocky mantle, preventing the circulation of nutrients and shutting down the planet-stabilizing geological cycles. Life, if it arose, might find itself starved of the essential elements needed to build and sustain itself. It is a classic case of having too much of a good thing. While life could potentially arise around hydrothermal vents on the icy "seafloor," it might struggle to become a truly global, planet-altering biosphere in the way it has on Earth.

At the opposite end of the planetary spectrum lies the desert planet. This is a world that is also in the habitable zone, but one that is water-starved. Its surface would be dominated by vast, arid landscapes of rock and sand, with liquid water confined to small, isolated oases, perhaps in polar ice caps, deep canyons, or small, shallow seas. Such a world might form from drier building blocks in the inner part of a protoplanetary disk, or it could be a world that once had more water but lost it over time, perhaps due to a weak magnetic field failing to protect its atmosphere from being stripped by its star. Mars is our solar system's own tragic example of this, a planet that was once warmer and wetter but has since devolved into a cold, global desert.

At first glance, a desert world might seem less promising for life than a water world. Yet, climate modelers have discovered that these arid planets have a surprising advantage: they possess a much wider effective habitable zone than a water-rich planet like Earth. The presence of large oceans makes a planet's climate exquisitely sensitive to changes in stellar brightness. As a star heats up, more water evaporates from the oceans, adding water vapor—a powerful greenhouse gas—to the atmosphere. This traps more heat, which causes more evaporation, creating a dangerous feedback loop that can lead to a runaway greenhouse effect, boiling the oceans away entirely. This is what sets the inner edge of the habitable zone for an Earth-like planet.

A desert planet, with its limited surface water, is far less susceptible to this fate. With little water to evaporate, it can orbit much closer to its star without triggering a runaway greenhouse effect. Similarly, a desert planet is more resistant to freezing. The vast expanses of dry land have a lower heat capacity than oceans and are also less reflective than ice, making it harder for the planet to fall into a global "snowball" state. This means the outer edge of the habitable zone also extends further out for a dry world. The Goldilocks zone, it turns out, is significantly more forgiving for planets that are a little parched.

The climate on a desert planet would be one of extremes. Without the moderating influence of large oceans, which absorb and release heat slowly, the temperature difference between day and night and between the equator and the poles would be dramatic. The atmosphere might be thinner and would certainly be drier, leading to clear skies and a surface exposed to more stellar radiation. Life on such a world would be a constant struggle for survival, likely clustered around the few stable sources of water. These oases would be the centers of biology, separated by vast, sterile expanses of desert, much like life in the most arid regions of our own planet.

This brings us to a compelling question: what is the ideal water budget for a habitable planet? Is there a "just right" amount of water, and does Earth represent that sweet spot? The answer seems to be a tentative yes. Having some exposed land and some oceans appears to offer the best of both worlds. The oceans provide a stable thermal reservoir and a medium for life, while the continents provide access to a rich supply of nutrients and participate in the planet-stabilizing climate cycles. The fraction of a planet's surface covered by land versus ocean is emerging as a potentially critical parameter for long-term habitability. A planet with too little land might struggle with nutrient supply, while a planet with too little ocean might lack the stability for complex life to evolve.

The balance is delicate. A planet's water budget has a profound influence on its entire geological and climatic evolution. The amount of water present can affect everything from the viscosity of the mantle, which influences plate tectonics, to the dominant cloud patterns in the atmosphere. It seems increasingly likely that for a planet to be truly Earth-like, it must not only be the right size and in the right orbit, but it must also have won the cosmic lottery and received something close to an Earth-like portion of water.

Distinguishing between these different types of worlds from light-years away is one of the great challenges for the next generation of telescopes. Our first clue, as always, comes from the planet's density. A rocky planet in the habitable zone with a conspicuously low density would be a prime candidate for a water world. Its large radius for its mass would suggest that a significant fraction of its volume is made up of low-density water rather than rock. Conversely, a planet with a density similar to Earth or Mars would be a better candidate for a terrestrial or desert world.

The most powerful tool for exploration, however, will be atmospheric spectroscopy. By analyzing the starlight that passes through or is reflected by a planet's atmosphere, we can search for the tell-tale spectral signature of water vapor. The abundance of water vapor in the atmosphere could provide a direct hint about the extent of the surface reservoir below. A planet with a very humid atmosphere, saturated with water vapor, would be a strong water world candidate. An atmosphere with only trace amounts of water vapor would point towards a desert planet. However, clouds can complicate this picture immensely, hiding the atmospheric gases below from our view.

Ultimately, the most definitive evidence will have to come from direct imaging. As telescopes become powerful enough to capture the faint light of a terrestrial exoplanet as a distinct point, we can watch how its brightness and color change as it rotates. A planet with oceans and continents would show a distinct rotational variability. As the planet turns, the different reflectivities of land, water, and clouds would cause its overall brightness to fluctuate in a repeating pattern.

An even more spectacular possibility is the detection of "glint." As a planet with an ocean rotates, there would be a moment when the ocean surface reflects the starlight directly towards our telescope, creating a brief, sharp flash of specular reflection, much like the glint of sunlight off a distant lake. The detection of such a glint would be unambiguous proof of a liquid surface, the holy grail in the search for an ocean on another world. A desert planet, by contrast, would show a more uniform, rocky reflection profile, while a water world would show no rotational color changes at all, though it might still exhibit glint. By combining all these techniques—density, spectroscopy, and long-term monitoring—we can hope to one day move beyond simple archetypes and begin to map the coasts of these distant, watery, and dusty worlds.

CHAPTER SIXTEEN: The Search for Biosignatures: Detecting Signs of Life

For all the astronomical distances and mind-bending physics involved in exoplanet science, the entire endeavor is propelled by a single, deeply human question: Are we alone? The discovery of thousands of worlds has transformed this question from philosophical speculation into a testable hypothesis. We have found planets of the right size in the right orbits, worlds that could plausibly host liquid water. The next, monumental step is to search for direct evidence of life itself. This is the search for biosignatures, the ultimate prize in the planet hunter's quest.

A biosignature is not a radio signal or a photograph of an alien city. It is any measurable property of a planet—be it a substance, an object, or a pattern—that provides scientific evidence of past or present life. We are not looking for organisms; we are looking for the fingerprints they leave on their world. Life, as we know it, is a messy and transformative chemical process. It takes in raw materials from its environment and expels waste products. Over geological timescales, a thriving global biosphere can fundamentally alter a planet's surface, its oceans, and most importantly, its atmosphere. It is this planetary-scale impact that we hope to detect from light-years away.

The search is therefore an exercise in cosmic forensics. We must scan the air of these distant worlds for tell-tale clues, for molecules that have no business being there in the quantities we find them. The core principle is the search for chemical disequilibrium. A dead planet, governed only by geology and chemistry, will eventually settle into a state of relative chemical equilibrium. Its atmosphere will be composed of gases that can happily coexist with each other and with the surface rocks. Life, however, is a relentless engine of disequilibrium. It constantly pumps out reactive gases that should not exist, creating an atmosphere that is in a perpetual state of chemical tension. Finding a world in such a state would be a profound discovery.

The most famous and sought-after of these atmospheric clues is molecular oxygen. On Earth, our atmosphere is composed of nearly twenty-one percent free oxygen, a quantity that is wildly out of place. Oxygen is an aggressive, highly reactive gas. It readily combines with other elements, a process known as oxidation. It rusts iron, reacts with volcanic gases, and oxidizes surface minerals. Without a constant and massive source, the free oxygen in our atmosphere would be scoured away in a geological blink of an eye. That source, of course, is life itself. Photosynthesis, the process used by plants, algae, and cyanobacteria, has been pumping oxygen into our atmosphere for over two billion years, maintaining it in a state of extreme disequilibrium.

To find an exoplanet with an atmosphere rich in oxygen would be a moment of historic significance. The spectral signature of oxygen, and especially its cousin molecule ozone (O_3), is strong and potentially detectable. Ozone is formed when high-energy sunlight splits oxygen molecules, and it has a particularly prominent absorption feature in the ultraviolet part of the spectrum. The presence of a significant ozone layer would be a powerful proxy for an oxygen-rich atmosphere below, and would immediately elevate a planet to the highest tier of interest.

However, nature is devious, and the universe has ways of creating false positives. An oxygen-rich atmosphere is not, by itself, a foolproof sign of life. Geochemists have imagined scenarios where non-biological processes could lead to a buildup of oxygen. One prominent theory involves water and light. On a planet orbiting close to its star, or one without a protective magnetic field, intense stellar radiation could split water vapor molecules in the upper atmosphere into hydrogen and oxygen. The lightweight hydrogen would easily escape into space, while the heavier oxygen atoms would be left behind, slowly accumulating over time. This process, known as photolysis followed by hydrogen escape, could potentially create a substantial oxygen atmosphere on a world that has lost its oceans and is now sterile. This is why context is everything. Finding oxygen on a wet, Earth-like world would be far more compelling than finding it on a hot, dry planet on the inner edge of its habitable zone.

While oxygen is the poster child for biosignatures, a more robust case can be made by looking for a combination of gases that should not coexist. Imagine finding an atmosphere that contains not only a large amount of oxygen but also a significant quantity of methane. Methane is the primary component of natural gas, and on Earth, the vast majority of it is produced by microbial life in environments ranging from cow stomachs to swamp mud. Crucially, methane and oxygen are chemically incompatible. In an oxygen-rich atmosphere, methane is quickly oxidized and destroyed. For both gases to be present in large quantities simultaneously, it implies that two separate, powerful sources are constantly replenishing them, fighting against the chemistry that seeks to eliminate them.

This simultaneous detection of an oxidizing gas (like oxygen) and a reducing gas (like methane) is the gold standard of atmospheric disequilibrium biosignatures. While geological processes can produce methane, and abiotic processes might produce oxygen, it is extraordinarily difficult to imagine a non-biological scenario that could produce massive quantities of both at the same time on the same world. Such a detection would be a powerful argument that the planet's atmosphere is being actively and continuously manipulated by biological processes. It would be the closest thing to catching the chemical breath of an alien biosphere.

The search extends beyond this classic pair. Scientists are investigating a whole suite of other molecules that could point to biological activity. One promising candidate is nitrous oxide (N_2O), sometimes known as laughing gas. On Earth, its atmospheric concentration is almost entirely due to the metabolic activity of bacteria as part of the nitrogen cycle. It has very few known abiotic sources, making it a potentially low-ambiguity biosignature. Its spectral features are located in a part of the infrared spectrum that is relatively clear of interference from other gases like water and carbon dioxide, making it a tantalizing target for telescopes like JWST.

Other suggestions include gases like ammonia, methyl chloride, and ethane. The challenge with many of these is that they can also be produced by geological or photochemical processes, muddying the waters. The key is not just to find a single gas, but to take a complete chemical inventory of the atmosphere. The relative abundances of all the detectable gases, when considered together, provide a much richer context than any single molecule. A detection will not be a simple check-box exercise but a complex puzzle, requiring scientists to build a holistic picture of the planet's atmospheric chemistry and ask if it makes more sense with or without the presence of life.

While the air of an exoplanet is the primary hunting ground, it is not the only place to look for clues. Life can also leave its mark on a planet's surface in a way that might be detectable from afar. On Earth, one of the most striking features of the land surface, when viewed in the right kind of light, is the "Vegetation Red Edge." Plant life, in its quest to perform photosynthesis, has evolved pigments like chlorophyll that are excellent at absorbing visible light, particularly in the red and blue parts of the spectrum. However, to avoid overheating, plants are also highly reflective in the near-infrared, a range of light just beyond what our eyes can see.

This results in a dramatic and sharp increase in reflectivity at the boundary between red visible light and near-infrared light. If you were to plot the spectrum of sunlight reflected from Earth, you would see this sharp cliff-like feature, the red edge, wherever there are large amounts of vegetation. Observing a similar, sharp spectral edge in the light reflected from a distant exoplanet would be a stunning discovery. It would suggest the presence of a widespread pigment on the planet's surface that is optimized for capturing energy from its star.

Of course, this too comes with caveats. An alien biosphere might evolve photosynthetic pigments that are entirely different from chlorophyll, tailored to the specific light spectrum of its parent star. A planet orbiting a cool red dwarf might have plants that appear black to our eyes, designed to absorb as much energy as possible across all wavelengths. This could create a "purple edge" or a "yellow edge," or no sharp edge at all. Furthermore, some minerals are known to have reflection features that could, from a great distance, mimic a biological edge. As with atmospheric signals, context and the sharpness of the feature would be critical in distinguishing a true sign of alien forests from a world of interesting rocks.

Biosignatures might also be temporal, changing with the rhythm of the seasons. On Earth, the concentration of carbon dioxide in the atmosphere follows a distinct annual cycle. It decreases during the spring and summer in the Northern Hemisphere as the vast forests and plankton blooms draw it out of the air for growth, and it increases in the fall and winter as that vegetation decays and releases it back. This sawtooth pattern in our planet's carbon dioxide levels is, quite literally, the signature of the biosphere breathing.

Observing a similar seasonal oscillation in the atmospheric gases of an exoplanet would be a powerful biosignature. It would imply the presence of a dynamic, planetary-scale process that is modulating the atmosphere in sync with the planet's orbit. To detect such a signal would require an extraordinary observational commitment, monitoring a single planet continuously for years to track the subtle changes in its air. While incredibly challenging, the payoff would be enormous, providing not just evidence for life, but a window into the metabolism of an entire world.

The immense difficulty of this search is compounded by the persistent problem of false negatives. We could be staring directly at an inhabited world and see nothing at all. Life might exist deep underground or in a subsurface ocean, like on Jupiter's moon Europa, leaving no discernible trace on the atmosphere above. The biosphere could be too sparse to produce a globally detectable signal. Or, most frustratingly, the planet could be shrouded in a thick, high-altitude haze or cloud deck that completely obscures our view of the atmosphere below, hiding its chemical secrets from our telescopes.

Ultimately, claiming the detection of a biosignature will not be a single event, but a long and careful process of elimination. It will begin with an observation of an anomaly—a gas that shouldn't be there, a strange reflection from the surface. Scientists will then have to work diligently to exclude every plausible non-biological explanation. This will involve building complex computer models of the planet's geology, atmosphere, and its interaction with its star, testing to see if any combination of abiotic factors could reproduce the observed signal.

Only when all known non-biological explanations have been exhausted and found wanting will the community begin to gain confidence in a biological origin. The first such announcement will be couched in cautious, probabilistic language. It will not be a declaration of "we have found life," but rather a statement that we have found something for which life is the leading hypothesis. It will be a discovery that demands further observation, confirmation by other teams using different instruments, and years of intense scrutiny and debate. It will be the beginning of a new conversation, one that shifts the focus from the search for worlds to the characterization of one truly special world, a pale blue or red or purple dot that is, for the first time, no longer just a planet, but a potential home.

CHAPTER SEVENTEEN: Alien Skies: Weather and Climate on Exoplanets

To gaze up at our own sky is to witness a familiar and dynamic ballet. We watch clouds of liquid water and ice drift and coalesce, driven by winds that carry the warmth of the sun from the equator to the poles. We experience the regular rhythm of the seasons, dictated by the gentle tilt of our planet's axis. This is weather, the daily atmospheric drama that has shaped our history, our art, and our understanding of the natural world. But the skies of other worlds dance to different tunes. The study of exometeorology, the weather and climate on planets beyond our solar system, has revealed a gallery of atmospheric behavior so extreme and so alien it challenges our very definition of a "stormy day."

The forecast for a Hot Jupiter, for instance, would be monotonous yet violent. These gas giants, tidally locked to their stars, exist in a state of permanent, lopsided heating. One hemisphere is forever baked by the relentless glare of its sun, while the other is trapped in an eternal, starless night. This immense temperature difference between the day and night sides is the engine that drives the most ferocious winds in the known universe. While the fastest winds on Earth might gust at a few hundred kilometers per hour, observations of Hot Jupiters have clocked winds moving at thousands of kilometers per hour, many times the speed of sound. On the exoplanet HD 189733b, astronomers measured winds blasting from the day side to the night side at over 8,600 kilometers per hour. On WASP-127b, an equatorial jet stream screams around the planet at a staggering 33,000 kilometers per hour.

This global, supersonic flow of gas has a fascinating and observable consequence. As the hot gas from the substellar point—the spot directly under the star—is whipped eastward by these winds, it carries its heat with it. The gas doesn't have enough time to radiate its energy away before it has been swept downstream. The result is a "hotspot offset." When we create thermal maps of these planets by observing their infrared glow as they orbit, the brightest, hottest point is not where it "should" be, directly facing the star. Instead, it is consistently found to be shifted to the east. This offset is a smoking gun for planetary-scale, super-rotating winds, a weather pattern that dominates the climate of these tidally locked giants. The James Webb Space Telescope (JWST) has produced detailed thermal maps of worlds like WASP-43b, clearly showing this offset and confirming a stark temperature difference of over 600 degrees Celsius between the hot, windy day side and the cooler, but still scorching, night side.

Of course, a sky is more than just wind. As this hot gas flows toward the cooler night side, it will eventually cool enough for some of its components to condense and form clouds. But the clouds of a Hot Jupiter are nothing like the puffy water clouds of Earth. They are clouds of rock and metal. On the very hottest of these worlds, with temperatures exceeding 2,000 degrees Celsius, the clouds may be composed of exotic, high-temperature minerals like aluminum oxide, the same substance as rubies and sapphires. Imagine a world where the haze is made of vaporized gemstones, condensing into liquid metal droplets on the cooler night side.

On slightly cooler, though still hellishly hot, gas giants, the primary cloud component is expected to be silicates. The day side is so hot that rock-forming minerals exist as vapor in the atmosphere. As the winds carry this rock vapor to the cooler night side, it condenses into tiny droplets of molten rock—liquid sand. These silicate clouds would then rain out of the atmosphere, falling into the deeper, hotter layers where they would vaporize again, completing a global, perpetual cycle of rock rain. Observations by JWST have directly detected the tell-tale spectral signature of these silicate clouds in the atmosphere of the "fluffy" planet WASP-107b, confirming a weather forecast of high clouds of sand. These clouds can be patchy, not a uniform blanket, and astronomers believe the nightsides of many Hot Jupiters are perpetually socked in with thick silicate cloud decks, while their daysides are often blasted clear by the intense stellar radiation.

This diversity of cloud material creates a kind of chemical weather sequence. Models predict that the composition of clouds on Hot Jupiters should change with temperature. The hottest worlds have high-level clouds of aluminum and titanium oxides. As you move to slightly cooler planets, those clouds form deeper down, and the dominant cloud layer becomes silicates. Cooler still, and even the silicate clouds form deeper, leaving a clearer upper atmosphere. At the "cool" end of the Hot Jupiter spectrum, with temperatures below about 1,250 degrees Fahrenheit, the skies may be dominated by a hydrocarbon smog, similar to that on Saturn's moon Titan.

While the weather on gas giants is a spectacle of extreme physics, the climate on rocky, potentially habitable worlds orbiting in the Goldilocks zone is of paramount importance to the search for life. Many of these planets, particularly those orbiting the common M-dwarf stars, are also expected to be tidally locked. This leads to the "eyeball planet" scenario, with a hot substellar desert, a frozen night side, and a potentially temperate ring of twilight in between. The weather and long-term climate of such a world depend critically on the properties of its atmosphere.

If the atmosphere is too thin, it cannot effectively transport heat from the day side to the night side. In this case, the night side would become so cold that the atmosphere itself would freeze and collapse onto the surface as ice, leaving the planet airless and sterile. However, if the atmosphere is thick enough, global wind patterns can be established that ferry enough heat to the night side to keep it from freezing out completely. The result is a world with permanent, planet-sized weather patterns. Intense convection would occur at the substellar point, where warm, moist air would rise and form a vast, permanent bank of clouds. High-altitude winds would then carry this moisture and cloud cover toward the night side, where the air would cool, sink, and flow back toward the day side along the surface.

For a rocky planet with an ocean, the situation becomes even more complex. The ocean itself can transport a huge amount of heat, far more efficiently than the atmosphere. Climate models suggest that oceanic currents could warm the night side so effectively that the entire planet remains ice-free, creating a world with a vast, open ocean on the day side and a potentially habitable liquid ocean on the perpetually dark night side. The presence and arrangement of continents further complicate the picture, channeling winds and ocean currents in ways that can dramatically alter a planet's climate. The seven-planet TRAPPIST-1 system provides a fascinating laboratory for these ideas. Climate models of these worlds suggest a range of possible climates, from a runaway greenhouse on the innermost planets to potentially habitable ocean worlds like TRAPPIST-1e, to frozen ice-worlds on the outer edge. JWST observations have already begun to test these models, suggesting, for instance, that TRAPPIST-1c, once thought to be a potential Venus-twin, likely has a very thin atmosphere or none at all.

Not all planets follow the neat pattern of a circular orbit. Many exoplanets travel on eccentric, elliptical paths that create extreme seasonal variations based not on axial tilt, but on distance from the star. A planet on such an orbit experiences a long, cold "winter" when it is far from its star, followed by a brief, torrid "summer" as it whips through its closest approach. This would drive dramatic, cyclical weather, with the atmosphere potentially freezing out during the long winter and then being violently re-vaporized and blown by powerful winds during the short, intense summer.

To understand these myriad possibilities, scientists rely on sophisticated computer simulations called General Circulation Models (GCMs). These are essentially the same complex programs used to forecast weather and model climate change on Earth, but adapted for the exotic physics of alien worlds. They solve the fundamental equations of fluid dynamics and radiative transfer, simulating the movement of heat and gas in a three-dimensional atmosphere. These models are crucial for interpreting our limited observations. For instance, a GCM can predict what the hotspot offset should be for a planet with a certain temperature and rotation rate, which can then be compared to actual data. They allow us to translate a simple light curve—a measure of a planet's brightness over time—into a two-dimensional map of its temperature, providing our very first weather maps of other worlds. This powerful synergy between observation and theory, between the photons gathered by telescopes and the intricate physics simulated in GCMs, is finally allowing us to look up at the alien skies and begin to understand the forecast.

CHAPTER EIGHTEEN: Star-Planet Interactions: The Dance of Gravity and Radiation

For much of our exploration so far, we have treated stars and their planets as separate entities in a rather one-sided relationship. The star, a colossal thermonuclear furnace, shines its light and energy outward, warming, shaping, and sometimes blasting the much smaller worlds that circle it. The planet, in this view, is a passive recipient, its fate dictated almost entirely by its star's whims and its own orbital position. This picture, while useful, is incomplete. A planetary system is not a monologue; it is a conversation, a continuous and intricate dance of gravity and energy. Planets talk back. They pull on their stars, they exchange magnetic fields, they can even alter their star's behavior and, in their final moments, change its very chemistry. This is the science of star-planet interactions, a field that explores the complex, two-way relationship that binds a sun to its worlds.

The most fundamental conversation between a star and a planet is held in the language of gravity, through the silent, powerful influence of tides. We are familiar with the tides on Earth, the daily rise and fall of the oceans pulled primarily by the Moon's gravity. The same force is at play in distant star systems, but often dialed up to an extreme degree. For a planet orbiting very close to its star, the difference in the star's gravitational pull on the planet's near side versus its far side is immense. This differential pull stretches the planet, deforming it from a perfect sphere into a slightly elongated, egg-like shape, known as a tidal bulge.

If the planet is rotating at a different rate than it revolves around its star, this bulge is constantly being dragged across the planet's surface, generating enormous internal friction. This process, known as tidal heating, converts the energy of motion into thermal energy, relentlessly warming the planet's interior from the inside out. As we've seen, this can be a powerful enough engine to maintain liquid water oceans on moons far from the Goldilocks zone. But for close-in planets, it is also a powerful brake. Over millions of years, this friction slows the planet's rotation until the drag ceases. This happens when the planet's rotation period exactly matches its orbital period, a state we know as tidal locking. One hemisphere becomes permanently locked facing the star, creating the worlds of eternal day and eternal night that are so common in the galaxy.

This tidal dance, however, does not just affect the planet's spin; it affects its orbit. The planet does not just raise a tidal bulge on itself; it raises a corresponding, though much smaller, bulge on the surface of its star. The interaction of these two bulges determines the long-term fate of the planet's orbit. For a planet like Earth, which orbits the Sun much faster than the Sun rotates, the tidal bulge on the Sun is pulled slightly ahead of the Earth's position. This forward bulge then pulls the Earth onward, giving it a tiny gravitational kick that causes it to slowly spiral outward, away from the Sun. The effect is minuscule, but over billions of years, it is measurable.

For a Hot Jupiter, the situation is reversed and far more dramatic. It orbits its star in just a few days, whipping around much faster than the star itself rotates. In this case, the tidal bulge raised on the star lags behind the planet. This lagging bulge of stellar material pulls backward on the planet, acting as a constant gravitational drag. It saps the planet's orbital energy, forcing it into a slow, inexorable death spiral. The planet gets closer and closer to its star, its year getting shorter and shorter, destined for an eventual fiery demise.

Astronomers are now actively hunting for this effect in action. One of the most famous candidates for orbital decay is WASP-12b, an ultra-hot Jupiter that orbits its star in just over one Earth day. Its orbit is so tight that it skims the upper atmosphere of its star. Observations have suggested that its orbital period is decreasing by a few milliseconds per year, consistent with the predictions of tidal decay. If these measurements hold true, WASP-12b has only a few million years left before it is torn apart and swallowed by its star.

The final act of this gravitational drama is governed by the Roche limit, a critical boundary around a star inside of which tidal forces are stronger than a planet's own self-gravity. Once a spiraling planet crosses this threshold, it cannot hold itself together. The star's gravity rips it apart, shredding it into a stream of gas and dust that forms a temporary ring around the star before being accreted onto its surface. This process of planetary engulfment is likely the final chapter for many of the Hot Jupiters we see today. Some may even have been consumed by their stars as they expanded in old age, a potential fate for the Earth when our own Sun becomes a red giant.

The conversation between a star and planet is not just gravitational; it is also profoundly energetic, mediated by the star's intense radiation and the constant outflow of charged particles known as the stellar wind. This interaction is most clearly seen in the process of atmospheric escape, which acts as a powerful sculptor of planets. A star's high-energy radiation, particularly in the extreme ultraviolet (XUV) and X-ray parts of the spectrum, can heat a planet's upper atmosphere to thousands of degrees. For a planet without a strong magnetic shield, this is a recipe for atmospheric erosion.

In a process called hydrodynamic escape, this intense heating can cause the upper atmosphere to "boil off" into space. The heated gas expands so rapidly that it flows away from the planet in a powerful outflow, dragging heavier elements along with the lighter hydrogen and helium. This is particularly effective for close-in planets orbiting young, active stars. The result is planetary transformation on a grand scale. A world that might have started its life as a Mini-Neptune, with a rocky core wrapped in a thick hydrogen blanket, can be stripped bare by its star over hundreds of millions of years. The stellar radiation sandblasts away its primordial atmosphere, leaving behind only the naked, rocky core.

This process of photoevaporation is thought to be responsible for one of the most striking features in the exoplanet census: the "evaporation valley" or "radius gap." When astronomers plot the distribution of planet sizes, they find a curious lack of planets with radii between about 1.5 and 2 times that of Earth. There are plenty of rocky Super-Earths below this size and plenty of gaseous Mini-Neptunes above it, but a mysterious valley in between. Photoevaporation provides a compelling explanation. Planets in this valley are in a transitional state. Either they are the exposed cores of Mini-Neptunes that have lost their atmospheres, or they are Super-Earths that just managed to hang on to their thinner secondary atmospheres. This valley is a fossil record of stellar radiation at work, a clear fingerprint of the star's ability to reshape its worlds.

We can see this process happening in real-time. Astronomers using the Hubble Space Telescope have observed several "evaporating" planets directly. The Hot Jupiter HD 209458b, famously nicknamed "Osiris," was found to be surrounded by a vast, teardrop-shaped envelope of escaping hydrogen, a colossal tail being blown off the planet by the stellar wind. Similarly, the "warm Neptune" GJ 436b has a gigantic hydrogen cloud, a testament to its atmosphere slowly bleeding away into space. These observations provide visceral proof that planets are not static objects, but are constantly being weathered and eroded by their parent stars.

This energetic conversation becomes even more complex when magnetic fields enter the picture. Just as Earth has a protective magnetic field, or magnetosphere, generated by its liquid iron core, many exoplanets are expected to have them as well. For a close-in planet, its magnetosphere can directly interact with the star's own magnetic field and the magnetized stellar wind. This Star-Planet Magnetic Interaction (SPMI) creates a dynamic and often violent relationship. The planet's magnetic field can act as a shield, deflecting the stellar wind and protecting the atmosphere. But it can also act as a funnel.

Magnetic field lines from the star and the planet can directly connect, creating a circuit between the two bodies. This allows charged particles from the stellar wind to be funneled directly down onto the planet's magnetic poles. This would generate aurorae of an intensity that would dwarf Earth's Northern and Southern Lights. Imagine a Hot Jupiter with aurorae that permanently encircle its poles, glowing with a power millions of times greater than our own, a constant light show driven by its connection to its star.

The search for these magnetic interactions is an active frontier. One tantalizing possibility is the detection of radio waves. In our own solar system, the powerful magnetic interaction between Jupiter and its volcanic moon Io generates intense radio emissions. Astronomers have been using radio telescopes to listen for similar signals from exoplanet systems, hoping to detect the tell-tale crackle of a planet's magnetosphere interacting with its star. While a definitive detection has remained elusive, a confirmed radio signal would be the first direct measurement of an exoplanet's magnetic field, a crucial property for assessing its ability to shield an atmosphere and potentially harbor life.

Remarkably, this magnetic connection is a two-way street. The planet doesn't just receive energy; it can also trigger activity on the star. The magnetic disturbance caused by a close-in planet plowing through the star's outer atmosphere, or corona, could be enough to induce a stellar flare or enhance the formation of starspots. Several studies have claimed to find evidence of this, noting that a star's flaring activity seems to be synchronized with the orbit of its Hot Jupiter. A flare might preferentially erupt when the planet is at a certain point in its orbit, suggesting the planet itself is the trigger.

This idea of planet-induced stellar activity is still debated, and disentangling it from a star's own natural variability is incredibly difficult. However, if confirmed, it would be a profound demonstration of the planet's influence. It would mean that a small, non-luminous world could reach out and poke its parent star, causing it to erupt in a brilliant flash of energy. The dance would be a true duet, with each partner influencing the other's behavior.

The most intimate and final form of star-planet interaction happens when a planet is completely consumed by its star. This can happen through the slow spiral of tidal decay or when a star expands into a red giant and engulfs its inner planets. When this occurs, the star's atmosphere is suddenly polluted with the heavy elements—the iron, silicon, and other "metals" (in astronomical terms)—that made up the rocky body of the planet. While a single Earth-sized planet might not make a huge difference to a Sun-like star, the ingestion of a gas giant or multiple terrestrial planets could leave a detectable chemical fingerprint in the star's atmosphere.

This has led astronomers to search for "polluted" stars, stars with unusually high abundances of elements like lithium. Lithium is an element that is typically destroyed over time in the hot interiors of stars, but it is preserved in planets. A star that shows an unexpectedly high level of lithium might be one that has recently ingested one of its own planets, dredging the planet's pristine lithium up to its surface.

This stellar forensics can provide clues about a planetary system's violent past. A system that has undergone a period of intense gravitational chaos, with planets being scattered and thrown into the star, might leave behind a star that is chemically enriched. The star's very composition becomes a tombstone, a record of the worlds it has devoured. This ultimate act of interaction, where the planet becomes part of its star, is a poignant reminder that in this cosmic dance, the partners are so intimately linked that they can, in the end, merge into one. The life of a planetary system is written not just in the orbits of its survivors, but in the very light of its central star.

CHAPTER NINETEEN: Rogue Planets: Worlds Without a Sun

In our cosmic imagination, the concept of a planet is inextricably tied to a star. We picture worlds bathed in the light of a sun, their seasons and years dictated by a stately orbit around a central point of gravity. Every planetary system we have explored, including our own, follows this familiar blueprint. Yet, the galaxy holds a hidden population of worlds that break this fundamental rule. These are the loneliest and most enigmatic of all planetary bodies: the rogue planets. They are worlds without a sun, gravitationally unbound to any star, adrift in the vast, cold darkness of interstellar space. They are the galaxy's orphans, nomads wandering the silent voids between the stellar islands.

The existence of these free-floating planets is a natural, if dramatic, consequence of the chaos that governs the birth of planetary systems. The prevailing theory for their origin is ejection. The early life of a solar system is a violent and unstable period, a gravitational game of billiards played with world-sized objects. As young, massive planets migrate and interact, their orbits can cross and become unstable. This can lead to a period of intense gravitational scattering where planets are flung about like stones from a slingshot. For every planet that settles into a stable orbit, another might be ejected from the system entirely, cast out into interstellar space to begin its solitary journey.

Computer simulations of planet formation suggest this is not a rare occurrence. In many models, the creation of a stable planetary system is an inefficient process that produces a significant number of castaways. This is particularly true in systems with multiple massive gas giants, whose powerful gravity can easily eject smaller, terrestrial-sized worlds. The number of rogue planets in the galaxy could therefore be a direct measure of the average chaos of planet formation. If systems form in an orderly fashion, rogues might be rare. If, as seems more likely, formation is a messy free-for-all, then the space between the stars could be littered with these exiled worlds.

A second, more sedate formation channel is also possible. Some rogue planets might never have had a home to be exiled from. It is conceivable that a small, dense clump of gas and dust within a vast molecular cloud could collapse under its own gravity, but lack the sufficient mass to ignite into a star. Instead of forming a star, it might form a gas giant-sized planet directly from the collapsing cloud, without ever having a stellar parent. This would be a process similar to the gravitational instability model of planet formation, but happening in isolation. These objects, born alone, would begin their existence as nomads. Differentiating between a planet that was ejected and one that was born a rogue is exceedingly difficult, but both pathways contribute to this unseen population.

Finding these solitary worlds is one of the greatest challenges in astronomy. They defy almost all of our standard detection methods. They do not orbit a star, so they neither transit a stellar disk nor induce a gravitational wobble. They have no nearby star to illuminate them, so they reflect virtually no light. They are fundamentally dark objects, almost perfectly camouflaged against the black backdrop of deep space. For decades, their existence was purely theoretical, a population of ghosts that scientists believed must be out there but had no way to see.

The key to unmasking these ghosts lies in using their gravity, not their light. As predicted by Einstein's general theory of relativity, any object with mass warps the fabric of spacetime around it. When a massive object passes directly in front of a more distant background star, its gravity acts like a lens, bending and magnifying the light of the source star. This phenomenon, gravitational microlensing, provides a unique and powerful tool for finding otherwise invisible objects.

As a rogue planet drifts across our line of sight to a distant star, it will briefly magnify the background star's light. An observer on Earth will see the star smoothly brighten and then fade back to its normal state over a period of time, producing a characteristic bell-shaped curve. The crucial clue to the nature of the lensing object is the duration of this event. A massive object like a star has a strong gravitational field and will produce a lensing event that lasts for weeks or even months. A small, planetary-mass object, however, has a much weaker gravitational field and will produce a much shorter event. The signature of a rogue planet is therefore a brief, sharp, isolated brightening of a background star, lasting anywhere from a few hours to a couple of days. This short duration is the smoking gun that tells us the lensing object is not a star, but a small, free-floating world.

This technique is a game of immense patience and sheer numbers. Microlensing events are incredibly rare, requiring a near-perfect alignment between the observer, the lens, and the distant source star. To catch one in the act, astronomers must monitor millions of stars, night after night, searching for these tell-tale brightenings. Large-scale surveys like the Optical Gravitational Lensing Experiment (OGLE) and the Microlensing Observations in Astrophysics (MOA) have been doing just this for years, staring at the dense, crowded star fields toward the center of our Milky Way galaxy.

It is through these patient vigils that the first evidence for rogue planets began to emerge. The surveys started detecting a number of very short-duration microlensing events, too short to be explained by stars or even brown dwarfs. In 2011, a landmark study based on MOA data sent shockwaves through the astronomical community. It analyzed the statistics of these short-duration events and came to a stunning conclusion: rogue planets were not just common; they might be overwhelmingly abundant, potentially outnumbering the stars in the Milky Way by at least a factor of two. The galaxy, it seemed, could be home to hundreds of billions, or even trillions, of these wandering worlds.

While subsequent studies have refined these numbers, with some suggesting a lower but still substantial population, the conclusion remains the same: the galaxy has a vast, hidden population of planets that belong to no solar system. The discoveries have become more specific and more spectacular. In 2020, the OGLE team announced the detection of OGLE-2016-BLG-1928, the shortest microlensing event ever observed, lasting just 42 minutes. The incredibly short timescale points to a lensing object with a mass smaller than Earth, possibly as small as Mars. This was the first strong evidence for a terrestrial-mass rogue planet, a tiny, rocky world drifting alone in the dark.

What would life be like on such a world? At first glance, the prospect seems bleak. A rogue planet receives no energy from a star. Its surface would be one of the coldest places in the universe, with temperatures plunging to near absolute zero. It would be a world of utter and perpetual darkness, save for the faint, diffuse glow of the distant Milky Way. Any atmosphere it once had would have long ago frozen solid and fallen to the ground as snow. By all conventional measures, it should be a sterile, lifeless iceball.

But the surface is not the whole story. As we have explored, a planet has its own internal engine, powered by the heat left over from its formation and the slow decay of radioactive elements in its core. For a sufficiently massive planet—a Super-Earth or a gas giant—this internal geothermal heating could be substantial. A thick atmosphere, particularly one rich in hydrogen, would act as a superb insulating blanket, trapping this internal heat and preventing it from leaking away into space.

This combination of a powerful internal furnace and a thick atmospheric blanket could create a remarkable possibility: a warm surface, with temperatures suitable for liquid water, even in the absence of a star. A rogue planet could be a self-sufficient habitable world. Life, if it existed, would not be based on photosynthesis. It would have to be chemosynthetic, deriving its energy from chemical reactions powered by the planet's internal heat, perhaps clustered around hydrothermal vents on the floor of a dark, globe-spanning ocean. Such a world would be a self-contained biosphere, its existence completely independent of any star, its inhabitants forever unaware of the universe beyond their atmospheric and icy shell. This radically expands the galactic real estate available for life, suggesting that habitable niches could exist in the vast, empty oceans of interstellar space.

The nature of these worlds would depend heavily on their origin and mass. A Jupiter-sized gas giant that was ejected from its system would be a dark, cold version of the planets we know, perhaps with its own family of tidally heated moons. A smaller, Earth-sized rogue would likely be too small to retain enough internal heat or a thick atmosphere, and would probably be a frozen, inert rock. The most intriguing candidates for rogue habitability are the Super-Earths, worlds massive enough to power their own geology and hold onto a thick insulating atmosphere for billions of years. These are the planets that could maintain liquid water on their surfaces through their own means.

The great challenge of studying rogue planets is that we get only a single, fleeting glimpse. A microlensing event is a one-time occurrence. Once the alignment passes, the planet vanishes back into the darkness, never to be seen again. We cannot go back to study it with other telescopes or watch it over time. We get a single data point that tells us its approximate mass, and that is all. This makes building a comprehensive picture of the rogue planet population incredibly difficult. We are like naturalists trying to study a new species of animal based on a single, blurry photograph taken at night.

This is all set to change. The next decade promises a revolution in the study of nomadic worlds, largely thanks to one upcoming mission: NASA's Nancy Grace Roman Space Telescope. Scheduled for launch in the mid-2020s, Roman is a wide-field observatory that will conduct a massive microlensing survey from space. Unhindered by the blurring effects of Earth's atmosphere and the interruption of the day-night cycle, Roman will stare at the central bulge of the Milky Way for years at a time, monitoring hundreds of millions of stars with unprecedented precision.

Its sensitivity and persistence will allow it to detect the short-duration microlensing events from rogue planets with unparalleled efficiency. It is expected to find hundreds, and possibly thousands,

117

of these wandering worlds, down to masses as small as Mars. For the first time, we will have not just a handful of tantalizing candidates, but a robust statistical sample. Roman will give us a definitive census of the galaxy's hidden population. It will tell us how many rogue planets there are for every star, what their typical masses are, and how they are distributed throughout the galaxy.

This data will provide crucial insights into the processes of planet formation. By comparing the number of rogue planets to the number of planets still bound to their stars, we can directly measure the efficiency—or inefficiency—of planetary system formation. A galaxy teeming with trillions of rogues would tell us that planetary ejection is a common and violent feature of solar system evolution. A galaxy with relatively few would suggest that most systems form in a more calm and orderly fashion. The ghosts of the galaxy are about to be counted, and their numbers will tell us a fundamental story about where planets come from, and how many are destined to wander alone through the great, open spaces of the cosmos.

CHAPTER TWENTY: Naming New Worlds: The Process of Cataloging Exoplanets

There is a uniquely human impulse to name things. We bestow names upon our children, our pets, our ships, the mountains we climb, and the streets we live on. A name transforms an object from an abstract entity into a place, a personality, a known quantity. It is an act of possession, of understanding, of connection. So, when humanity began discovering planets beyond our own solar system in earnest, a fundamental question arose, one that had little to do with physics or chemistry: What do we call them? Who gets to name a new world? The answer, it turns out, is a fascinating story of scientific necessity, bureaucratic order, and a newfound desire to connect the public with these distant, alien shores.

To the casual observer, the names given to most exoplanets are a bewildering and distinctly unromantic jumble of letters and numbers. Names like HD 189733 b, Kepler-186f, or OGLE-2016-BLG-1195Lb do not exactly roll off the tongue or inspire poetic verse. They sound less like new worlds and more like serial numbers, Wi-Fi passwords, or designations for military hardware. This is, in a way, precisely the point. The primary purpose of an exoplanet's name is not to be beautiful, but to be functional. It is a scientific designation designed for clarity, precision, and unambiguous identification within vast astronomical databases.

The official body tasked with overseeing this cosmic census is the International Astronomical Union, or IAU. Founded in 1919, the IAU is the internationally recognized authority for assigning designations to celestial bodies. Its role is to prevent a state of absolute chaos where a single planet might have dozens of different names in different countries or by different research groups. They are the librarians of the cosmos, ensuring every new discovery is cataloged according to a clear and logical system.

Without this central authority, collaborative science would grind to a halt, buried under a mountain of confusing and contradictory names.

The system the IAU has established for exoplanets is an extension of the one used for multiple-star systems. It is simple, logical, and scalable. The first part of a planet's name is simply the name of the star it orbits. This is usually not a poetic name like Sirius or Polaris, but a designation from a major astronomical catalog. For instance, "HD 189733" refers to the star's entry number in the Henry Draper Catalogue, a massive spectral classification of over a quarter of a million stars compiled in the early 20th century. This catalog name immediately tells an astronomer a great deal about the star, including its approximate location and type.

The second part of the name is a lowercase letter, starting with 'b'. The first planet discovered orbiting a given star receives the designation 'b', the second is named 'c', the third 'd', and so on. The star itself is considered the implicit 'a' component of the system. This lettering system reveals a crucial piece of information: the order of discovery, not the order of the planets from the star. The planet closest to the star might be named 'd', while the one farthest out could be 'b'. This is a practical necessity, as planets are not always found in a neat sequence from innermost to outermost. A small, inner planet might be discovered years after a large, obvious outer planet.

This system ensures that names do not need to be changed or shuffled around every time a new world is added to a known system. For example, in the famous TRAPPIST-1 system, the seven known planets are named TRAPPIST-1b, c, d, e, f, g, and h, all in the order of their discovery. This method provides a stable and expandable framework for the tens of thousands of planets that may one day be found. It is not romantic, but it is scientifically robust, a vital tool for organizing the rapidly growing galactic menagerie.

As planet-hunting surveys have become more sophisticated, the catalog designations of the parent stars have become a kind of honor roll for the missions themselves. The name "Kepler-186f" tells you that this planet orbits the 186th star found to have planets by the Kepler Space Telescope. A name beginning with "CoRoT" belongs to a discovery by the French-led CoRoT mission. "WASP" denotes a planet found by the Wide Angle Search for Planets, while "HAT" signifies a discovery by the Hungarian Automated Telescope Network. These prefixes act as a form of scientific provenance, instantly telling researchers which team and which instrument made the initial find.

More recently, the Transiting Exoplanet Survey Satellite (TESS) has introduced its own layer to this process. When TESS spots a potential transit, the host star is given a TESS Object of Interest (TOI) number, such as TOI 700. This is a candidate, a person of interest in a cosmic investigation. If subsequent observations by other telescopes confirm that the transit is indeed caused by a planet, that planet is officially cataloged and named, in this case becoming TOI 700d. This two-stage process highlights the rigorous verification that must occur before a world gets its official name. It must first graduate from being an "Object of Interest" to a confirmed and cataloged planet.

For decades, these alphanumeric designations were the beginning and end of the story. While scientifically necessary, many in the astronomical community and the wider public felt something was missing. The name 51 Pegasi b commemorates a scientific achievement, but it doesn't capture the wonder of discovering a new world. This disconnect became particularly apparent with the rise of the internet, which brought these discoveries to a global audience. The public, fascinated by these new worlds, wanted names they could connect with, names that had meaning and history.

This growing sentiment led to a brief and slightly bizarre period where private companies saw a commercial opportunity. Several organizations began offering unofficial services to "buy" a name for an exoplanet. For a fee, anyone could name a distant world after a loved one, a favorite celebrity, or even themselves, and receive a handsome but entirely unofficial certificate. The IAU was forced to step in, issuing firm and repeated statements that such naming services had no official standing. The naming rights for celestial bodies, they argued, are not a commodity to be bought and sold. They belong to all of humanity.

Recognizing the genuine public desire for more inspiring names, the IAU decided to meet the public halfway. Instead of allowing a free-for-all, they initiated a series of carefully organized public naming campaigns called NameExoWorlds. These campaigns provide a structured and inclusive way for people around the world to participate in the naming process. The first major campaign, launched in 2015, assigned a selection of well-studied exoplanets and their host stars to astronomical clubs and organizations worldwide, who were invited to submit name proposals based on specific themes. The public then voted on the winning names.

The results were a resounding success, injecting a rich tapestry of mythology, literature, and culture into the exoplanet catalog. The very first exoplanet discovered around a Sun-like star, 51 Pegasi b, was given the proper name Dimidium, a Latin word for "half," referencing its half-Jupiter mass. Its star was named Helvetios, after a Celtic tribe that once inhabited Switzerland, honoring the Swiss astronomers who discovered it. The star 55 Cancri, home to five known planets, was named Copernicus, and its planets were named Galileo, Brahe, Lippershey, Janssen, and Harriot, creating a pantheon of Renaissance astronomers.

Perhaps the most poignant names were given to the very first exoplanets ever confirmed, the three small, rocky worlds orbiting the pulsar PSR B1257+12. This pulsar, the spinning corpse of a dead star, was named Lich, after a powerful undead creature in fantasy fiction. Its planets were named Draugr, Poltergeist, and Phobetor, different names for undead or evil spirits from Norse mythology and Greek mythology. These names perfectly capture the eerie, almost unnatural feeling of finding planets orbiting the ghost of a star.

A subsequent NameExoWorlds campaign in 2019 took the concept even further, assigning a star-planet system to over 110 different nations, allowing each country to run its own national campaign to name its designated world. This resulted in a stunningly diverse collection of names drawn from global cultures. Ireland named its star and planet after the mythological dogs Bran and Tuiren. Jordan chose names for a star and planet from the ancient, rose-red city of Petra. Malaysia named its system after the gemstone Merdelima. This initiative brilliantly transformed the abstract act of naming a distant point of light into a celebration of global heritage and a point of national pride.

The guidelines for these public names are straightforward and sensible. Names should be pronounceable, non-offensive, and not named after living individuals. They must not be purely commercial in nature, and they should ideally come from themes of cultural, historical, or geographical significance. The process has provided a wonderful solution to the naming dilemma, creating a dual system that serves two different but equally important purposes. For the working scientist, the precise, searchable alphanumeric designation remains the primary identifier. For the public, and for the sheer romance of exploration, a growing number of worlds now have proper names, names that tell a story.

This cataloging process extends beyond simply assigning a name. It involves maintaining a definitive and publicly accessible record of a planet's properties. The primary repository for this information is the NASA Exoplanet Archive, a massive online database that serves as the official census for the field. For every confirmed planet, the archive lists its key parameters: its orbital period, its mass, its radius, its density, its discovery method, and links to the original scientific papers. This archive is an indispensable tool, allowing scientists to study the exoplanet population as a whole, to search for trends, and to identify promising targets for future study.

The process of naming a world is, in the end, a reflection of our relationship with the cosmos. The systematic, data-driven cataloging represents our scientific impulse to understand, to classify, and to build a coherent picture of the universe based on evidence. The more recent tradition of assigning culturally significant names represents our human impulse to connect, to tell stories, and to find our own place among the stars. One system speaks to the mind, the other to the heart. Together, they allow us to take the growing list of distant worlds and transform them from mere data points into a true atlas of the galaxy, with each new name adding another destination to a map that is only just beginning to be drawn.

CHAPTER TWENTY-ONE: The Kepler Mission: A Revolution in Planet Hunting

Before 2009, the study of exoplanets was largely a collector's game. Astronomers, using the radial velocity method and early transit surveys, hunted for worlds one by one, celebrating each new discovery as a hard-won prize. The catalog of known planets grew steadily, but it was a skewed collection, heavily biased toward the large, close-in Hot Jupiters that were easiest to find. The biggest and most profound questions remained unanswered. Were planets rare or common? Were systems like our own the norm or an oddity? And the ultimate question that drove the field: How many Earth-like worlds, orbiting in the temperate zones of their stars, were really out there? Answering these questions required a radical new approach, a shift from discovery to demographics. It required a census. It required a telescope designed for a single, audacious purpose: to count the planets. That telescope was Kepler.

NASA's Kepler Space Telescope was not built to be a jack-of-all-trades observatory like Hubble. It was a highly specialized instrument, a planet-hunting machine honed for one specific task: to measure the brightness of stars with unprecedented precision. Its scientific heart was a single instrument, a photometer, which was essentially a giant digital camera. At its core was a mosaic of 42 charge-coupled devices (CCDs), which together formed a 95-megapixel detector, the largest camera ever launched into space at the time. This enormous digital eye was placed at the focus of a 0.95-meter telescope, a design known as a Schmidt camera, which gave it an exceptionally wide and sharp field of view.

Unlike Hubble, which orbits the Earth, Kepler was placed into a unique, Earth-trailing heliocentric orbit. It was programmed to slowly drift away from the Earth as both circled the Sun. This clever orbit was crucial for its mission. It freed the telescope from the obscuring effects of Earth's atmosphere, eliminated the stray light and thermal variations of an Earth orbit, and allowed it to point continuously at a single spot in the sky without the Earth ever getting in its way. It was a profoundly lonely and stable vantage point from which to conduct a patient, unwavering vigil.

Kepler's mission strategy was as simple as it was ambitious. It would stare, and stare, and stare. For four straight years, it fixed its gaze upon a single, pre-selected patch of the sky, a 115-square-degree field located in the northern constellations of Cygnus and Lyra, along the Orion arm of the Milky Way. This particular patch was chosen for several reasons. It was densely packed with a typical mix of stars, offering a rich hunting ground. Critically, it was positioned well above the plane of our solar system, ensuring that the Sun, Earth, and Moon would never drift through its field of view, allowing for nearly uninterrupted observation.

The mission's goal was to detect the tiny, periodic dimming of a star's light that occurs when an orbiting planet passes in front of it—a transit. To find a true Earth analogue, a planet in a one-year orbit, Kepler needed to see at least three transits to confirm the planet's existence and establish its period. This meant it had to stare at the same group of stars, without blinking, for a minimum of three to four years. For ninety-six percent of that time, Kepler did just that, continuously monitoring the brightness of more than 150,000 stars every thirty minutes. It was the most extensive and intensive stellar survey ever undertaken.

The torrent of data that flowed back from the spacecraft was immense. Every day, the telescope gathered more raw data than the entire Hubble Space Telescope had collected in its first two decades of operation. Turning this river of pixels into a list of planetary candidates was a monumental task for the science team on the ground. The first step was to produce a light curve for every single star, a graph of its brightness over time. These curves were not clean, flat lines. They were noisy, filled with instrumental artifacts and the star's own natural variability, such as starspots and flares.

Sophisticated software pipelines were developed to clean this raw data, correcting for instrumental drift and filtering out the astrophysical noise. Once the data was cleaned, automated algorithms searched through the light curves, hunting for the characteristic U-shaped dip of a planetary transit. Any star showing a promising, repeating transit-like signal was flagged and given a new designation: a Kepler Object of Interest, or KOI. This was the master list of potential planets, the starting point for a rigorous process of vetting and follow-up observations by ground-based telescopes to weed out false positives and confirm the planetary nature of the candidate.

The first results from Kepler, announced in 2010, were exciting but not entirely surprising. The initial batch of confirmed planets, including Kepler-4b through Kepler-8b, were mostly Hot Jupiters. These were the "low-hanging fruit" of the survey. Their large size produced a deep transit dip, and their short orbital periods meant they transited frequently, making them easy to spot and confirm in the early data. These initial discoveries were crucial for verifying that the telescope and the analysis pipeline were working as expected. They were the warm-up act for the main event.

Soon after, the true revolution began. In January 2011, the Kepler team announced the discovery of Kepler-10b, the mission's first confirmed rocky planet. It was a "lava world," smaller than two Earth radii but orbiting so close to its star that its surface was likely molten. It was not habitable, but it was definitive proof that Kepler could find planets far smaller than Jupiter. A few months later came Kepler-11, a stunning discovery that revealed a complete planetary system unlike any other seen before. It featured no fewer than six planets, five of which were huddled in orbits smaller than Mercury's, a remarkably compact and flat system that hinted at a much calmer formation history than our own.

As the mission continued and the data analysis matured, the trickle of discoveries became a flood. The sheer number of planetary candidates Kepler uncovered was staggering. By the end of its primary mission, it had identified thousands of KOIs, a number that dwarfed the total count of all previously known exoplanets. More importantly, the statistical nature of the survey allowed scientists to answer the big questions that had motivated the mission in the first place. The results were transformative.

First and foremost, Kepler proved that planets are not the exception, but the rule. The data indicated that, on average, there is at least one planet for every star in our galaxy. The cosmos was not empty, but teeming with worlds. Second, and perhaps most significantly, it completely overturned our understanding of planetary demographics. While previous surveys had been dominated by gas giants, Kepler revealed that this was simply an observational bias. In reality, small planets are far more common than large ones. The galaxy, it turns out, prefers to build Super-Earths and Earth-sized worlds, the very class of planet our own solar system seems to be missing.

This led directly to the mission's holy grail: an estimate for the number of potentially habitable, Earth-like worlds. This value, known to astronomers as eta-Earth ($\eta_\oplus$), represents the fraction of Sun-like stars that host a planet between about one and two times the size of Earth, orbiting within the star's habitable zone. After years of careful analysis, the answer emerged from the Kepler data. The numbers vary depending on the exact definitions used, but the conclusion was unmistakable: eta-Earth was not a tiny fraction. Conservative estimates suggested that as many as one in five Sun-like stars might harbor such a world. When extrapolated across the entire Milky Way, this implied the existence of billions of potentially habitable, rocky planets. They were no longer a theoretical fantasy; they were a statistical certainty.

Kepler also opened our eyes to the sheer architectural variety of planetary systems. It found the first definitive examples of circumbinary planets, worlds that orbit two stars at once. The discovery of Kepler-16b in 2011 confirmed that the fictional double sunset on Tatooine was a physical reality in our galaxy. Its light curve was a complex pattern of dips, showing a primary transit across one star, a secondary transit across the other, and the eclipses of the two stars by each other. Kepler's continuous, long-term monitoring was perfectly suited to disentangling these complex signals, and it went on to find a dozen such "Tatooine" worlds.

Then, in May of 2013, four years into its mission, disaster struck. The second of Kepler's four gyroscopic reaction wheels, the devices used to aim the telescope with exquisite precision, failed. With only two working wheels, Kepler could no longer hold its steady gaze on its original field. For many, it seemed like the mission was over, a brilliant success cut short. But the ingenuity of the NASA engineering team gave the old telescope a second life. They devised a brilliant workaround for the broken wheel, a new mission concept called K2.

The solution was to use the pressure of sunlight itself as a "virtual" third reaction wheel. By carefully orienting the spacecraft so that the solar wind pushed evenly on its solar panels, engineers could balance the telescope and achieve a level of stability sufficient for planet hunting. This meant Kepler could no longer stare at a single spot indefinitely. Instead, it would point at different fields along the ecliptic plane—the path of the Sun and planets across our sky—for shorter observational campaigns of about 80 days each.

The K2 mission, which began in 2014, was a resounding success. While its shorter campaigns meant it could not find planets with year-long orbits, it surveyed a much larger and more diverse area of the sky than the original mission had. Over the next four years, K2 observed hundreds of thousands of new stars, including many more nearby and bright red dwarfs, young stars in stellar nurseries, and even other galaxies and supernovae. It discovered hundreds more confirmed planets and thousands of new candidates. A key benefit of K2 was that it found many transiting planets orbiting stars that were much brighter than those in the original Kepler field. These bright-star systems were ideal targets for follow-up studies by other telescopes, providing a rich trove of worlds for the James Webb Space Telescope to characterize.

Finally, on October 30, 2018, after nine years of revolutionary service, the Kepler Space Telescope ran out of the fuel needed to maintain its orientation and was officially retired. It had discovered more than 2,600 confirmed planets, with thousands more candidates still awaiting confirmation. But its true legacy is not just in the numbers. Kepler single-handedly transformed exoplanet science from a discipline of discovery into a statistical field. It gave us our first real planetary census, proving that small, potentially rocky worlds are common throughout the galaxy. It demonstrated, with hard data, that the raw materials for another Earth are anything but rare. It took a question that had been asked for millennia—are there other worlds like ours?—and for the first time in human history, it gave us a number.

CHAPTER TWENTY-TWO: The James Webb Space Telescope: A New Era of Exoplanet Science

For years, the James Webb Space Telescope was an astronomer's dream, a promise of a new vision so powerful it would rewrite textbooks. Its journey from concept to reality was a long and arduous one, marked by technical hurdles, budget overruns, and moments of profound doubt. Finally, on Christmas Day 2021, the culmination of decades of work by thousands of scientists and engineers was launched into the void. As it traveled to its distant outpost a million miles from Earth and unfurled its magnificent, gold-plated mirror, a new era in astronomy began. While built to be a versatile time machine capable of peering back to the dawn of the cosmos, one of its primary and most anticipated roles was to turn its golden eye upon the worlds orbiting other stars. If the Kepler mission was the great planet hunter that conducted a census, JWST is the great planet characterizer, a forensic laboratory designed to dissect alien atmospheres.

JWST is not simply a bigger version of the Hubble Space Telescope. It is a fundamentally different kind of observatory, designed from the ground up to master a different kind of light. Its power lies in the infrared. While Hubble sees the universe primarily in the visible and ultraviolet light our eyes are sensitive to, JWST is optimized to capture the longer wavelengths of infrared radiation. This focus is not accidental; it is the key to unlocking the secrets of exoplanets. The chemical fingerprints of the most interesting molecules for life—water, methane, carbon dioxide—are most prominent in the infrared part of the spectrum. An observatory blind to this light would be blind to the most telling clues in an exoplanet's atmosphere.

Everything about JWST's design serves this purpose. Its massive 6.5-meter primary mirror, composed of eighteen hexagonal, beryllium segments coated in a microscopically thin layer of gold, is a near-perfect collector of infrared light. To see the faint infrared glow of a distant planet, the telescope itself must be incredibly cold; otherwise, its own heat would swamp the signal. This is why JWST is not in Earth orbit. It is stationed at the second Lagrange point, or L2, a gravitationally stable spot 1.5 million kilometers away. Here, it can use its enormous, five-layer sunshield—a technological marvel the size of a tennis court—to block the light and heat from the Sun, Earth, and Moon simultaneously, allowing the telescope to cool to a frigid minus 233 degrees Celsius. This cold, stable vantage point gives JWST the breathtaking sensitivity needed for its mission.

The area where JWST has had its most immediate and revolutionary impact is in the field of transmission spectroscopy. This technique, which analyzes the starlight filtering through a planet's atmosphere during a transit, has been used by Hubble for years, but JWST has taken it to an entirely new level. Its combination of a huge mirror and powerful spectrographs allows it to capture transmission spectra with a clarity and level of detail that was previously unimaginable. Almost immediately after it began science operations, it delivered a result that showcased this power.

In August 2022, astronomers announced that JWST had made the first-ever definitive detection of carbon dioxide in the atmosphere of an exoplanet. The target was WASP-39b, a "hot Saturn" orbiting a Sun-like star some 700 light-years away. Previous telescopes had found hints of various gases, but the signal from carbon dioxide, a crucial tracer of a planet's formation history and metal content, remained elusive. JWST's Near-Infrared Spectrograph (NIRSpec) did not just hint; it delivered an unambiguous, mountain-like peak in the spectrum right where carbon dioxide was predicted to be. It was a watershed moment, moving the field from tentative suggestions to confident chemical inventories.

The full spectrum of WASP-39b, combining data from multiple JWST instruments, went even further. It revealed not just a single molecule, but a detailed portrait of a complex and dynamic atmosphere. The data confirmed the presence of water vapor, sodium, and potassium, but also found the surprising signature of sulfur dioxide. This molecule is not expected to exist in chemical equilibrium in such an atmosphere. Its presence was a smoking gun for photochemistry—the process by which high-energy starlight breaks apart other molecules, which then recombine to form new ones. For the first time, we were seeing not just the static ingredients of an alien atmosphere, but direct evidence of active chemical reactions being driven by its star.

JWST has also pushed the boundaries of emission spectroscopy, the study of the thermal glow from a planet's own heat. As a planet passes behind its star in a secondary eclipse, JWST can measure the minute drop in infrared light, isolating the planet's contribution. This provides a direct measurement of the planet's temperature. The first major target for this technique was the TRAPPIST-1 system, a famous nearby system of seven Earth-sized rocky planets orbiting a cool red dwarf star.

In early 2023, an international team announced that JWST had measured the dayside temperature of the innermost planet, TRAPPIST-1b. It found a temperature of about 230 degrees Celsius. This might seem academic, but it was a crucial clue. A planet with a thick, circulating atmosphere would have distributed heat from its day side to its night side, resulting in a cooler overall dayside temperature. The high temperature measured by JWST strongly suggested that the planet has little to no atmosphere at all. A subsequent measurement of the second planet, TRAPPIST-1c, found a similar result, indicating it too was likely an airless rock. These were pivotal, if somewhat sobering, results. They marked the first time we had been able to probe the atmospheric conditions of any planet as small and as cool as the rocky worlds in our solar system, and provided our first real check on which of these famous habitable-zone-adjacent planets might actually be habitable.

The ultimate challenge for exoplanet science has always been direct imaging, the attempt to capture an actual picture of a planet by separating its faint light from the blinding glare of its star. Here too, JWST is a game-changer. It is equipped with several state-of-the-art coronagraphs, tiny masks that can be positioned to block the starlight with exquisite precision. Its stability in space, free from the blurring of Earth's atmosphere, gives it a tremendous advantage over even the largest ground-based telescopes.

In September 2022, the telescope took its first direct image of an exoplanet, a gas giant named HIP 65426 b. This planet is about seven times the mass of Jupiter and orbits its star at a distance roughly one hundred times greater than the Earth-Sun distance. Its great mass, youth, and wide separation made it a relatively easy target, one that had been imaged before from the ground. But JWST was able to see the planet in multiple infrared wavelengths that are inaccessible from Earth, providing new detail. While not the discovery of a new world, it was a spectacular proof of concept. It demonstrated that JWST's coronagraphs worked as designed and that the observatory was ready to begin its quest to image new and unknown worlds. That moment came in 2025, with the announcement of TWA 7 b, the first exoplanet discovered via direct imaging with JWST—a world with a mass comparable to Saturn's.

The true power of directly imaging with JWST, however, lies not just in taking a picture, but in what can be done with that planet's light once it is isolated. That light can be passed into a spectrograph to reveal the planet's chemical composition. This opens up a whole new population of planets to atmospheric study: those that do not transit. It allows us to analyze the atmospheres of the gas giants in wide orbits, the true analogues to our own Jupiter and Saturn, whose properties hold vital clues about the formation of planetary systems as a whole.

JWST's reach extends beyond mature planets to the very nurseries where they are born. With its powerful infrared vision, it can peer deep into the dusty protoplanetary disks that surround young stars. It can resolve these disks with stunning clarity, revealing the gaps and spiral structures being carved out by unseen, newly forming planets. It can also perform a chemical census of these disks, identifying the raw materials available for planet formation, including water and complex organic molecules. For the first time, scientists can connect the dots, studying the chemical inventory of a planet-forming disk and comparing it to the atmospheric composition of a mature planet, tracing the journey of the chemical building blocks of life from nebula to world.

The telescope is not without its own challenges. Studying planets orbiting active red dwarf stars like TRAPPIST-1 is complicated by the star's own variability. Flares and starspots on the star's surface can create signals that contaminate the faint light from the planet's atmosphere, requiring complex and careful analysis to disentangle the two. The sheer volume and complexity of the data require new and more sophisticated atmospheric models and retrieval techniques to interpret.

Every spectrum from JWST is a puzzle, a complex interplay of chemistry, physics, and geology. But it is a puzzle we are now equipped to solve. The James Webb Space Telescope has moved exoplanet science from an era of discovery to an era of characterization. The goal is no longer just to find new dots on a map, but to understand what those dots are made of. We have entered the age of comparative planetology on a galactic scale, and for the first time, we have a tool powerful enough to begin the search for the chemical signs of life in the skies of other worlds.

CHAPTER TWENTY-THREE: The Future of Exoplanet Exploration: Upcoming Missions and Telescopes

The discoveries of the Kepler mission and the James Webb Space Telescope have been nothing short of revolutionary. One gave us the planetary census, proving that worlds are everywhere; the other gave us the forensic tools to dissect their skies. Yet, for all their power, these trailblazers have also illuminated the vast territory that still lies beyond our reach. We have found thousands of planets, but we have only just begun to understand them. The next chapter in this great exploration will be written by a new generation of observatories, both on the ground and in space, each designed to push the frontiers of discovery further and to tackle the most profound questions that remain. The future of exoplanet exploration is a story of bigger mirrors, sharper eyes, and more audacious goals, all aimed at moving from a catalog of worlds to a deep understanding of them, and perhaps, to finally capturing a portrait of another Earth.

On the ground, the next leap forward is taking the form of a new class of behemoths: the Extremely Large Telescopes, or ELTs. For decades, the largest optical telescopes in the world have hovered around the 8-to-10-meter mark. The ELTs will shatter that barrier, with primary mirrors so vast they will collect more light than all other professional telescopes in history combined. This jump in light-gathering power and resolution will open new windows into exoplanet science.

Leading the charge is the appropriately named Extremely Large Telescope, a project of the European Southern Observatory (ESO). Being constructed atop Cerro Armazones in the hyper-arid Atacama Desert of Chile, the ELT will be the largest optical and infrared telescope in the world. Its primary mirror is a staggering 39.3 meters (129 feet) in diameter, composed of 798 hexagonal segments working in concert. This colossal eye will be able to perform feats that are currently impossible. For exoplanet science, its goals are twofold. First, it will be a supreme machine for atmospheric characterization. By gathering the faint light of a transiting planet's atmosphere, the ELT's high-resolution spectrographs will be able to dissect the chemical makeup of worlds with unprecedented detail, including smaller, rockier super-Earths. Second, it is designed to be a formidable direct imaging instrument. Paired with advanced adaptive optics to cancel out atmospheric blur and sophisticated coronagraphs to block starlight, the ELT will push to directly image planets closer to their stars and smaller in mass than ever before. It aims to take the first direct images of mature, Jupiter-like planets in nearby systems and perhaps even image the glow of large, young, rocky worlds.

Not far from the ELT, another giant is rising in the Chilean desert: the Giant Magellan Telescope (GMT). Taking a different design approach, the GMT will use seven of the largest single mirrors ever fabricated, each 8.4 meters across, to create a single observing surface with an effective aperture of 25.4 meters. Its unique configuration will provide exceptionally sharp images. Like the ELT, the GMT will be a powerhouse for both transmission spectroscopy and high-contrast direct imaging, enabling detailed studies of exoplanet atmospheres and pushing the boundaries of what worlds we can see directly. Its instruments are being designed specifically to probe the environments of nearby rocky planets, searching for the molecular building blocks of life.

The third member of this new triumvirate is the Thirty Meter Telescope (TMT), planned for the summit of Mauna Kea in Hawaii. Following a segmented-mirror design similar to the iconic Keck telescopes but on a much grander scale, the TMT's 30-meter aperture will give it extraordinary light-gathering power and resolution in the Northern Hemisphere. The project has faced significant hurdles, navigating the complex cultural and environmental sensitivities of its proposed location, a site considered sacred by many Native Hawaiians. Should it proceed, its scientific potential for exoplanet research is immense. It would work in concert with its southern counterparts, the ELT and GMT, to provide all-sky coverage, allowing astronomers to study targets anywhere in the sky. It would be particularly powerful for characterizing the many planets found by the Kepler mission, which stared at a patch of the northern sky. Together, these three ground-based giants will form a global network, a new generation of eyes on the universe that will transform our ability to study other worlds from the surface of our own.

While the ELTs prepare to conquer the ground, a new fleet of specialized probes is being readied for space. The next great flagship mission on the launchpad is NASA's Nancy Grace Roman Space Telescope. Roman is a true powerhouse, a wide-field observatory with a 2.4-meter mirror—the same size as Hubble's, but with a field of view one hundred times larger. This vast cosmic panorama makes it a survey instrument of unmatched capability. Its primary mission is to conduct a massive gravitational microlensing survey. As we have seen, microlensing is uniquely sensitive to planets in wide orbits, far from their stars, and to worlds that have no star at all. Roman will stare at the dense star fields of the galactic bulge, detecting the brief flashes of magnified light that signal the presence of these unseen worlds. It is predicted to discover thousands of new exoplanets, completing the planetary census begun by Kepler. It will provide the first robust statistics on the population of "cold Jupiters," the true analogues to the gas giants in our own solar system, and will give us a definitive count of the lonely rogue planets wandering the galaxy.

But Roman has a second, equally revolutionary instrument on board: the Coronagraph Instrument. This is a technology demonstrator, a camera packed with cutting-edge starlight-suppression technologies, including advanced masks and deformable mirrors. Its purpose is to test and refine the techniques needed for ultra-high-contrast direct imaging in space. While the main microlensing survey will find planets indirectly, the Coronagraph Instrument will be pointed at nearby, bright stars in an effort to take actual pictures of their known giant planets. Its goal is to achieve a contrast a thousand times better than any previous space-based coronagraph, pushing into a new regime where it could image giant planets in the light they reflect from their star, not just their own internal heat. The lessons learned from Roman's coronagraph will be the essential training ground for future missions that aim to directly image an Earth.

Following close behind Roman are two key missions from the European Space Agency (ESA) designed to answer specific, targeted questions. The first is PLATO, the PLAnetary Transits and Oscillations of stars mission. PLATO is a planet hunter, but with a very specific goal: to find Earth-like planets in Earth-like orbits around Sun-like stars. Unlike Kepler, which stared deep into a small patch of sky, or TESS, which scans the whole sky quickly, PLATO will employ a "long-stare" strategy on a very wide field of bright, nearby stars. By monitoring these stars for years at a time, it is designed specifically to detect the rare, shallow transits of rocky worlds with orbital periods of a year or more. PLATO's other key strength is asteroseismology. By studying the subtle, rhythmic changes in a star's brightness, it can measure the star's fundamental properties—its age, mass, and radius—with incredible precision. This is vital, because to understand a planet, you must first understand its star. By finding Earth-sized worlds and precisely characterizing their suns, PLATO will provide the best-vetted list yet of potentially habitable Earth-twins.

The second ESA mission, ARIEL (Atmospheric Remote-sensing Infrared Exoplanet Large-survey), is not designed to find new planets at all. It is a dedicated characterization mission, a specialist in atmospheric chemistry. ARIEL's mission is to perform the first large-scale, uniform survey of exoplanet atmospheres. It will observe a diverse sample of about one thousand known transiting planets, from scorching gas giants to rocky super-Earths. For each one, it will obtain a high-quality spectrum, measuring the chemical fingerprints of the gases in its atmosphere. The power of ARIEL is not in studying one planet, but in studying many. By creating this large, unbiased dataset, it will allow scientists to perform comparative exoplanetology on a grand scale. They can look for patterns, asking how a planet's atmospheric composition relates to its mass, its temperature, or the type of star it orbits. ARIEL will provide the crucial data needed to test our theories of how planets form and evolve, and to understand the full diversity of planetary climates in the galaxy.

These near-term missions are all stepping stones toward a single, ultimate goal, a mission concept so ambitious it will likely define astronomical research for the middle of the 21st century: the Habitable Worlds Observatory (HWO). Envisioned as the next great flagship mission to follow Hubble and JWST, HWO has one primary, breathtaking objective: to directly image and characterize a population of potentially habitable Earth-like planets orbiting nearby Sun-like stars. This is the quest to find a "pale blue dot," a true Earth twin, and to search its atmosphere for the chemical signs of life.

The technological challenge is immense. To see the faint reflected light of an Earth-like planet next to its star, a telescope needs to achieve a contrast ratio of about ten billion to one. HWO is envisioned as a large space telescope, with a mirror perhaps six meters in diameter, operating in the ultraviolet, visible, and near-infrared light where an Earth-twin would be brightest. It would be equipped with an extraordinarily advanced internal coronagraph, building on the technology pioneered by Roman. An alternative and complementary technology is the starshade, a separate, large spacecraft shaped like a giant sunflower, that would fly in precise formation tens of thousands of kilometers away from the telescope. The starshade would be positioned to cast a perfect, deep shadow of the target star over the telescope, allowing the faint light of the planet to shine through.

With the planet's light isolated, HWO would then perform the most anticipated observation in the history of science. It would pass that light into a spectrograph and search for biosignatures. It would look for the tell-tale spectral lines of water vapor, suggesting oceans below. It would hunt for ozone, the proxy for a dense, oxygen-rich atmosphere. And it would search for the chemical disequilibrium of gases like methane coexisting with oxygen, the most robust sign we know of a living, breathing biosphere. HWO is still a concept, a grand challenge laid down for the next generation of scientists and engineers, likely to launch in the 2040s. But it represents the culmination of our search, the point at which we transition from finding potential habitats to directly assessing them for the presence of life.

Beyond these flagships, a host of other innovative ideas are percolating in research labs. Some scientists are looking into large-scale space-based interferometers, arrays of smaller telescopes flying in formation that can combine their light to achieve the resolution of a single, much larger telescope, powerful enough to map the surfaces of nearby worlds. Others are dreaming of an even more futuristic concept: using the Sun itself as a giant gravitational lens. By sending a small probe to the Sun's gravitational focus point, a region starting at about 550 times the Earth-Sun distance, it would be theoretically possible to use the Sun's immense gravity to magnify the light of a distant exoplanet so powerfully that one could map its continents and oceans. The engineering challenges are formidable, but the payoff would be the ultimate close-up. From the giant telescopes rising in our deserts to the ambitious probes being planned for deep space, the tools are being forged for the next phase of our journey, a phase that promises to bring these other worlds, and the possibility of life upon them, into sharper focus than ever before.

CHAPTER TWENTY-FOUR: The Human Factor: The Possibility of Interstellar Travel to Exoplanets

For as long as we have known of other worlds, we have dreamed of visiting them. The discovery of thousands of exoplanets has transformed this dream from a vague fantasy into a concrete, if distant, ambition. We now have addresses for these worlds, detailed maps of their solar systems, and tantalizing clues about their skies and surfaces. We have identified specific, tangible destinations, some of which may even be habitable. Yet, the gap between knowing a place exists and being able to stand upon it is the largest and most challenging gulf humanity has ever contemplated crossing. The journey to even the nearest exoplanet is a voyage not of thousands, but of trillions of kilometers, a challenge that pushes the very limits of physics, engineering, and human endurance.

The sheer scale of interstellar space is a concept that our minds, evolved to deal with terrestrial distances, struggle to comprehend. The standard unit of measurement, the light-year, is itself an abstraction of an immense distance—the nearly 9.5 trillion kilometers that light travels in a single year. Our nearest exoplanetary neighbor, Proxima Centauri b, is just over four light-years away. This sounds deceptively close, but it is a journey of about forty trillion kilometers. To put this in perspective, the Voyager 1 spacecraft, which has been traveling away from Earth for nearly half a century and is the fastest and most distant human-made object, would take approximately 75,000 years to cover that distance. A journey to a world a few hundred light-years away, a mere stone's throw in galactic terms, would take our current technology millions of years. The stars are not just far away; they are far away on a scale that renders all of our conventional modes of travel utterly insignificant.

The fundamental obstacle that stands between us and the stars is not just a matter of engineering better engines; it is a law of physics known as the Tsiolkovsky rocket equation. First derived by the Russian spaceflight pioneer Konstantin Tsiolkovsky in 1903, this elegant formula governs the motion of all rockets. It reveals a brutal truth: a rocket's final change in velocity depends on the speed of its exhaust and the ratio of its initial mass (full of fuel) to its final mass (empty). To go faster, you need a faster exhaust, or you need to carry a lot more fuel relative to the mass of your ship. This creates a vicious cycle of diminishing returns. To reach interstellar speeds, you need a colossal amount of fuel. But that fuel has mass, and to accelerate that extra mass, you need even more fuel. The result is that for a conventional chemical rocket, like the ones that took humanity to the Moon, the amount of fuel required for an interstellar journey would be many times the mass of all the oceans on Earth. The rocket equation is a tyrant, and it dictates that we cannot get to the stars by simply building bigger fuel tanks.

To escape this tyranny, we need to abandon chemical rockets and explore entirely new forms of propulsion, technologies that can achieve far higher exhaust velocities or sidestep the need to carry fuel at all. The first and most plausible step up from chemical rockets is nuclear propulsion. One well-studied concept is the nuclear thermal rocket. Instead of a chemical reaction, a compact nuclear fission reactor is used to heat a propellant, such as liquid hydrogen, to extreme temperatures, expelling it through a nozzle. The exhaust velocity is much higher than in a chemical rocket, making the system two to three times more efficient. This technology was extensively developed in the United States under Project Rover in the 1960s and is considered a relatively near-term possibility for rapid travel within our own solar system. For interstellar travel, however, it would still result in journey times of many thousands of years.

A far more radical and powerful concept from the atomic age was Project Orion, a design for a nuclear pulse propulsion vehicle. The idea was as simple as it was terrifying: the spacecraft would carry a large number of small nuclear bombs and release them one by one out of its rear. Each bomb would detonate a short distance behind the ship, and the resulting plasma blast would slam into a massive, heavily shielded "pusher plate" at the back of the vessel. The ship would ride these successive shockwaves, lurching forward with each pulse. While the thought of riding a wave of atomic explosions is startling, the physics is sound. The exhaust velocity of a nuclear blast is immense, and detailed studies suggested that an Orion-style ship could theoretically reach speeds of up to five percent of the speed of light, potentially making the trip to Proxima Centauri in under a century.

Looking further ahead, scientists dream of harnessing the power of nuclear fusion, the same process that fuels the stars themselves. A fusion rocket would use magnetic fields to contain and compress isotopes of hydrogen until they fuse into helium, releasing enormous amounts of energy. This energy would heat a propellant or be directed as exhaust, achieving speeds far greater than any fission-based design. The British Interplanetary Society conducted a detailed study for such a craft in the 1970s, called Project Daedalus. Their design was for a massive, two-stage, uncrewed probe that could reach Barnard's Star in about fifty years. The primary obstacle, of course, is that we have not yet managed to build a controlled fusion reactor on Earth that produces more energy than it consumes. Harnessing this power for a rocket remains a challenge for the distant future.

An even more elegant solution to the rocket equation is to leave the fuel at home. Beamed-energy propulsion concepts aim to push a spacecraft from a distance, freeing it from the burden of carrying its own fuel and engine mass. The most prominent of these ideas is the light sail. The craft would consist of a vast, ultra-thin, highly reflective sail, perhaps kilometers across but weighing only a few grams. A powerful laser array, either on Earth or in orbit, would fire a concentrated beam of light at the sail. The constant pressure from the photons in the laser beam would gently but continuously accelerate the sail. With a powerful enough laser and a light enough sail, theoretical speeds could reach twenty percent of the speed of light or more.

The Breakthrough Starshot initiative, announced in 2016, is a real-world research program dedicated to developing this very technology. Its goal is to send a fleet of tiny, gram-scale "nanocrafts"—essentially a camera, a sensor, and a transmitter on a chip—attached to small light sails on a flyby mission to Proxima Centauri. The journey would take about twenty years. The challenges are still immense. They include building a 100-gigawatt laser array, keeping the beam focused on a tiny sail across billions of kilometers, and figuring out how to slow down at the destination, a problem for which there is currently no good solution for a simple sail.

The absolute pinnacle of propulsion, according to our current understanding of physics, would be a matter-antimatter engine. When a particle of matter meets its corresponding antiparticle, they annihilate each other completely, converting one hundred percent of their mass into energy, as described by Einstein's famous equation, $E=mc^2$. This is the most energy-dense reaction known to be possible. A spacecraft powered by antimatter annihilation could theoretically reach speeds in excess of ninety percent of the speed of light. The problem is that antimatter is staggeringly difficult and expensive to produce and even harder to store. It cannot touch any normal matter without annihilating. We can currently produce antimatter only in minuscule quantities in particle accelerators, and the amount needed for an interstellar trip would require many millennia of production at our current rates.

Beyond these concepts, we enter the realm of speculative physics, ideas that manipulate spacetime itself. The Alcubierre drive, or warp drive, is a theoretical solution to the equations of general relativity that proposes a way to travel faster than light without violating causality. It does not involve the ship itself moving fast, but rather creating a "warp bubble" of spacetime that contracts space in front of the ship and expands it behind. The ship would ride this wave of spacetime like a surfer, potentially covering vast distances in a short amount of time. The catch is a significant one: creating such a bubble appears to require the existence of "exotic matter" with negative mass, a substance we have never observed and which may not be physically possible.

Even if we were to solve the immense challenge of propulsion, the ship is only half the problem. The human crew would face a series of biological and psychological hurdles that are just as daunting. For journeys at a significant fraction of the speed of light, the strange effects of special relativity come into play. Time itself would pass differently for the travelers than for those they left behind on Earth. This phenomenon, known as time dilation, means that for a crew on a ship traveling at ninety-nine percent the speed of light, a journey to a star fifty light-years away might take them only a few years of their own time. But when they returned, they would find that over a century had passed on Earth. They would be adrift in time as well as in space.

For slower, sub-light journeys that might take centuries or millennia, the only way for humans to make the trip would be aboard a generation ship. This would not be a vehicle in the traditional sense, but a self-contained, mobile world. It would have to be a perfectly closed-loop ecosystem, capable of recycling one hundred percent of its air, water, and waste, and growing its own food for hundreds of generations. The engineering challenge of creating such a perfect, failure-proof biosphere is colossal. Even more profound are the sociological and psychological challenges. How would a society develop in complete isolation, confined to an artificial world for its entire history? What would be the ethics of launching a journey that the initial crew would never see completed, condemning their descendants to a life and a destiny they did not choose?

A common trope in science fiction is to bypass this problem with suspended animation or hibernation. While some animals can hibernate for a season, putting a human into a state of long-term stasis is currently far beyond our medical capabilities. The biological processes involved in safely lowering a human's metabolism for centuries and then reviving them without cellular damage or neurological decay are completely unknown.

Perhaps the most persistent and deadly hazard of an interstellar journey is radiation. The space between the stars is not empty; it is permeated by a constant flux of galactic cosmic rays, high-energy particles accelerated to near-light speed by distant supernovae. Earth's magnetic field and atmosphere provide a powerful shield against this radiation. A spaceship, however, would be fully exposed. Over a centuries-long journey, the cumulative radiation dose to the crew would be lethal, dramatically increasing cancer risks and causing damage to the central nervous system. Providing adequate shielding is a major mass problem; a thick enough shield of water, rock, or lead would make the ship too heavy to accelerate. Lighter, more exotic materials or powerful magnetic shields are being investigated, but the problem remains a formidable one.

Given these seemingly insurmountable obstacles for human travel, the most realistic first steps toward the stars will almost certainly be robotic. Small, intelligent, and unburdened by the fragile needs of human biology, robotic probes can endure the long journey times and harsh radiation. Missions like the aforementioned Breakthrough Starshot, which aim to send wafer-sized probes to our nearest neighbors, represent the logical vanguard of interstellar exploration. These tiny emissaries, powered by light sails, could make the journey in a single human generation, performing a flyby to take pictures, measure the planet's properties, and search for an atmosphere. They would be our scouts, our eyes and ears in another solar system, gathering the crucial data needed to determine if a destination is truly worth the monumental effort of a human journey. They could answer the fundamental question of whether a planet is habitable long before we develop the means to inhabit it ourselves.

The possibility of interstellar travel forces us to confront our own limitations and our most profound motivations. The challenges are so great, the timescales so long, that any such project would require a commitment of resources and a continuity of purpose across many generations, far beyond anything humanity has ever attempted. It is a dream that lies at the very edge of possibility, a testament to our species' innate and perhaps irrational desire to see what lies over the next hill, even when that hill is another sun, trillions of kilometers away.

CHAPTER TWENTY-FIVE: Unanswered Questions: The Great Mysteries of Exoplanet Science

For all that we have learned in a few short decades, the story of exoplanets is one that is still being written, its greatest chapters yet to come. Each new discovery, each detailed spectrum, seems to answer one question while posing a dozen more. We have lifted the veil on a galaxy teeming with worlds, but in doing so, we have revealed the profound depth of our own ignorance. We are no longer asking *if* planets exist, but *why* they exist in such bewildering variety, and what this variety means for the ultimate question of life. The field of exoplanet science is now defined by its great mysteries, the grand, unanswered questions that will drive exploration for generations to come.

At the very heart of the discipline lies the planet formation puzzle. Kepler taught us that the most common type of planet in the galaxy is a world we do not have: the super-Earth or mini-Neptune. This raises a fundamental question: Why is our solar system so unusual? Or, to put it another way, why is the galaxy's most popular model of planet missing from our own lineup? The line between a large, rocky super-Earth and a small, gassy mini-Neptune appears to be one of the most important boundaries in nature, separating potentially habitable worlds from uninhabitable gas dwarfs. The "radius valley," a curious scarcity of planets between 1.5 and 2 Earth radii, suggests that planets are born as one type or the other, or are transformed by their stars. We see evidence that stellar radiation can boil away the atmospheres of mini-Neptunes, carving them down into barren super-Earth cores. But is this the whole story? Or do they form in fundamentally different ways from the start? Recent models suggest super-Earths may form in the dry, inner regions of a planetary disk, while mini-Neptunes form from icy material beyond the snow line, creating two distinct families of planets from the outset. Untangling these competing theories is crucial to understanding the prevalence of truly Earth-like worlds.

The origin of Hot Jupiters remains another persistent enigma. We know these behemoths must have migrated from their distant birthplaces, but how? Was it a gentle, inward spiral through the primordial gas disk, a process that might leave smaller, inner planets intact? Or was it a violent game of gravitational billiards, a chaotic scattering event that ejected other planets from the system and left the Hot Jupiter as the sole, scorched survivor? The alignment of their orbits provides clues; many Hot Jupiters are found on tilted, even backward, orbits relative to their star's spin, a strong hint that a violent, chaotic past is the more likely culprit. Yet we also find systems with Hot Jupiters that have smaller companion planets, suggesting a calmer history. It seems nature has more than one way to make a monster.

This leads to the broader mystery of planetary architecture. The neat, orderly layout of our solar system—small, rocky planets inside; large, gas giants outside—appears to be a relative rarity. Instead, the galaxy is filled with compact systems of super-Earths in tight, resonant orbits, planets so close their years are measured in days or weeks. Other systems are dominated by lonely, eccentric giants. Why is there such a zoo of outcomes? It seems the final arrangement of a system is a sensitive function of its initial conditions—the mass of its protoplanetary disk, the chemical composition of its star, and the chaotic dance of migration and instability. Understanding the rules that govern this process is essential if we are to understand our own place in the cosmic menagerie.

Even when we find a planet of the right size in the right place, we are confronted by the profound mystery of what lies within. For a super-Earth in the habitable zone, knowing its mass and radius is not enough. The data can be maddeningly degenerate, meaning multiple interior structures can produce the same result. A planet with a given density could be a barren, rocky world with a massive iron core, or it could be a water world with a smaller core, drowned beneath a globe-spanning ocean hundreds of kilometers deep. Breaking this degeneracy is a critical task for future observatories.

A related and equally important unknown is the prevalence of geological activity on these worlds. On Earth, plate tectonics acts as a planetary thermostat, recycling materials and regulating our climate over geological time. It is often considered a prerequisite for long-term habitability. Do super-Earths, with their higher mass and stronger gravity, have more vigorous plate tectonics? Or does that same gravity create a "stagnant lid" crust that shuts down this vital process? Recent studies looking at Earth's own early history suggest that mobile plate tectonics may not have been active when life first arose, indicating it might not be a requirement for life's origin, even if it is for its long-term success. But we do not know if Earth's history is a universal template. Observing these distant worlds is our only hope of finding out whether active geology is a common feature of rocky planets.

Just as mysterious is the question of planetary magnetism. A robust magnetic field, like Earth's, is a vital shield, protecting a planet's atmosphere from being stripped away by the stellar wind and deflecting deadly cosmic rays from the surface. Without it, any hope for surface life is likely doomed. Yet, we have no unambiguous measurements of the magnetic field of any exoplanet. The race is on to find one. Astronomers are using radio telescopes to search for the tell-tale signals produced when a planet's magnetosphere interacts with its star, a technique that has so far set upper limits but not yet yielded a definitive detection. Other, more subtle methods propose looking for the polarization of light passing through an atmosphere as it is warped by a magnetic field. Discovering how common planetary magnetic fields are, and how their strength relates to a planet's size and rotation, is a fundamental and unanswered question in the study of habitability.

The atmospheres of these worlds present their own set of profound puzzles. One of the most frustrating but important is the cloud conundrum. With increasing frequency, JWST is delivering spectra that are maddeningly flat and featureless. This is the signature of a high-altitude cloud or haze layer that acts like a shroud, completely obscuring our view of the atmosphere below. It prevents us from taking a chemical inventory and determining the planet's nature. What are these clouds made of? How do they form and at what altitude? Why are some planets clear while others are perpetually overcast? Understanding the microphysics of alien clouds—whether they are made of silicate dust, exotic salts, or photochemical smog—is a crucial and complex challenge that scientists are now tackling with coupled cloud-haze models. Cracking the cloud problem is essential if we are to peer into the skies of potentially habitable worlds.

Nowhere is this challenge more apparent than in the study of the TRAPPIST-1 system. These seven Earth-sized worlds, three of which lie in the habitable zone, represent our best natural laboratory for studying temperate rocky planets. Yet, initial observations with JWST have yielded a series of perplexing and sometimes contradictory results. Early measurements of the two innermost planets suggested they were bare, airless rocks. More recent, detailed observations of the innermost planet, TRAPPIST-1b, now present a paradox: the data is consistent with either an airless, volcanically active surface of unweathered rock, or a thick, hazy atmosphere rich in carbon dioxide. This ambiguity highlights the immense challenge of characterizing these small worlds and the intricate interplay between a planet's surface, its atmosphere, and the activity of its star. Answering the question of whether any of the TRAPPIST-1 planets are truly habitable will require years of further observation and more sophisticated models.

These atmospheric studies may eventually help us solve a mystery closer to home: the origin of Earth's own air. Our atmosphere is secondary, formed long after the planet itself. But what was the dominant source? Was it volcanic outgassing from the interior, or was it delivered by a barrage of water-and-gas-rich comets from the outer solar system? Exoplanetary systems provide us with the only possible comparison. By studying the atmospheric compositions of other rocky worlds of different ages and in different environments, we can begin to see patterns, distinguishing between worlds with outgassed atmospheres and those whose air seems to have been delivered from outside. These distant worlds are a mirror, reflecting possible histories for our own planet.

All of these questions culminate in the final, ultimate mystery: Are we alone? The search for biosignatures is now underway, but it is fraught with ambiguity. Even our most promising ideas are based on life as we know it. Are we suffering from a profound failure of imagination? Is the search for oxygen and methane a form of chemical chauvinism, blinding us to the possibility of "weird life" that follows completely different rules? The concept of Hycean worlds—cool planets with global oceans under hydrogen-rich atmospheres—has greatly expanded the search space for life, suggesting that habitable conditions could exist on planets very different from Earth. However, even this idea is debated, with some models suggesting these worlds would be too hot to maintain their oceans. Finding life may require us to think beyond familiar chemistry and search for biosignatures we have not yet conceived of.

There is also the question of whether we should be looking for biology or technology. The search for biosignatures looks for the planetary-scale impact of any life, microbial or otherwise. The search for technosignatures, such as industrial pollution in an atmosphere or the heat signature of a planet-spanning civilization, narrows the focus to intelligent, technology-using life. Technosignatures might be less ambiguous than biosignatures, but they are also presumed to be much rarer. Which search is more likely to succeed first is a matter of intense and philosophical debate.

This leads, inexorably, to the final and most profound question of all, one first posed in a casual lunchtime conversation and now looming over the entire field of astrophysics. With billions of potentially habitable planets now thought to exist in our galaxy alone, many of which are billions of years older than Earth, why is the cosmos so quiet? This is the Fermi Paradox, the eerie and unsettling silence in a galaxy that should be humming with life. Perhaps life is incredibly rare. Perhaps intelligence is a one-in-a-trillion evolutionary fluke. Perhaps civilizations inevitably destroy themselves. Or perhaps they are out there, and we are simply not listening in the right way. The paradox sharpens with every new world we discover. Each new potentially habitable planet is another voice that *could* be speaking, deepening the mystery of the great silence. The search for exoplanets is, in the end, a search for our own context. We do not yet know if our solar system is a bustling cosmic metropolis, a lonely rural outpost, or a unique and solitary miracle. Answering that question is the grand adventure that lies before us.

Printed in Dunstable, United Kingdom

71966915R00087